AF328682

DESIGN
ENGINEERING
REFOCUSED

MIX
Paper from
responsible sources
FSC
www.fsc.org
FSC® C015829

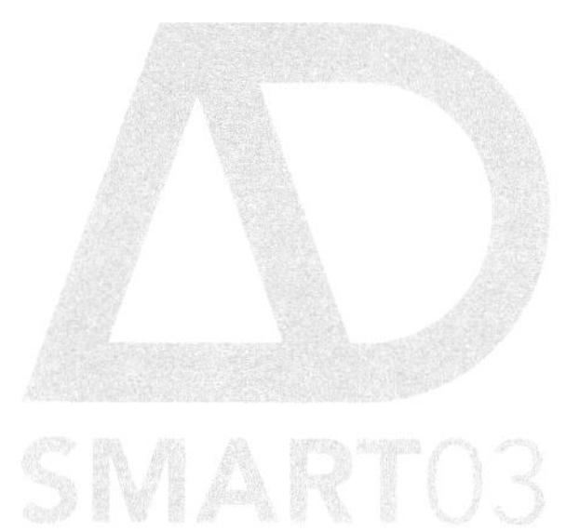

DESIGN ENGINEERING REFOCUSED

WILEY

Hanif Kara and Daniel Bosia

CONTENTS

ACKNOWLEDGEMENTS

We would first like to thank the directors and all the staff at AKT II, past and present; without them and their projects, this publication would not have been possible. It also goes without saying that this applies to all their clients, patrons and collaborators over the many years, as without them there would be no projects. We would also like to thank Professor John Ochsendorf for a considered and most welcome foreword. Hanif would like to single out Dean Mohsen Mostafavi at GSD for inspiring, advising and supporting this publication and for his poignant contribution 'Future Focus' at the start of this book.

To the authors who provided us with this book's exceptional content; Jordan Brandt, Marco Cerini, Diego Cervera de la Rosa, Philip Isaac, Jeroen Janssen, Sawako Kaijima, James Kingman, Alessandro Margnelli, Panagiotis Michalatos, Ed Moseley, Richard Parker, Andrew Ruck, Adiam Sertzu, Djordje Stojanovic, Edoardo Tibuzzi, Martijn Veltkamp and Marc Zanchetta.

Thanks, in particular, to Harvard University GSD and the AA, for encouraging us, but also to all the other institutions over the years.

We would like to thank Joshua Simpson and Kate Hobson for supervisory editing, Jan Friedlein, Erica Choi and Fritzie Manoy for graphic design, and Jessica Wainwright-Pearce for all the support in coordinating both internally and with the Wiley team.

Finally, we wish to thank our families for putting up with the late nights and long weekends during the construction of this book over the last two years.

FUTURE FOCUS MOHSEN MOSTAFAVI

In architecture, the connection between the logic of a form and the logic of its structure always used to be thought of as direct, linear, and overtly rational. Right up to the latter part of the twentieth century, the principle of upright structural support, represented by vertical columns and horizontal beams, provided the dominant method for the conceptualisation and design of most buildings.

This Cartesian mode of imagining the reciprocities between form and structure, in all its many iterations, is of course still very much with us today. It continues to be the reference point for the vast majority of contemporary architectural projects, shaping our imaginations as well as the prevalent methods of the building industry, which in turn feed back into the design process through, for example, the considerations of cost and period of construction.

Buildings produced through a column grid structure can vary enormously in their systematic adherence to the relation between form and structure. But this relation was itself radically transformed during the second part of the twentieth century, with the evolution of concrete thin shell structures that brought about a synthetic unity between form and structure. Engineers such as Pier Luigi Nervi, Eduardo Torroja, and Felix Candela were instrumental in developing forms that were no longer purely reliant on traditional methods of building construction. In place of structure as form, they proposed the notion of form as structure.

Through its exploration of both the geometric properties of shell structures and the elastic qualities of reinforced concrete, the work of these engineers produced a radically different conception of architectural form. Their research resulted in spatial forms that at times seem to closely resemble shapes and patterns found in nature.

These developments in the field of engineering also have some parallels with the earlier work of the Scottish biologist and mathematician D'Arcy Wentworth Thompson, whose classic book *On Growth and Form*, first published in 1917, would become a primary source for subsequent studies of morphogenesis—the idea of forms and their connections with plants and animals. Similarly, one key consideration of the work presented in this book is the shift from linear to non-linear geometry. The structural behaviour of many contemporary designs no longer follows—or perhaps more importantly, necessarily needs to follow—traditional methods for calculating structural forces. In addition, technological advances have made it possible to both imagine and construct forms that previously would have been nearly impossible to conceive.

While often focused on the articulation of continuous skins and variations in the curvature of building envelopes, these explorations can nevertheless also be utilised to transform our traditional conceptions of architectural design and construction.

It is against this backdrop, and with advances made in computation, materials and fabrication procedures, that the contributions to this book have taken shape. *Design Engineering Refocused* proposes a new way of considering the hybrid relationship between design and engineering. For it is in the space of entanglement and reciprocities between these two types of practice that the authors have discovered innovative ideas and unexpected solutions that respond to typical programs and everyday needs of users and clients.

Mohsen Mostafavi is Dean of the Harvard Graduate School of Design and the Alexander and Victoria Wiley Professor of Design.

ENGINEERING AS EXPLORATION JOHN OCHSENDORF

In his 2004 essay 'In Search of Brunel',[1] architect Charles Correa lamented the hyper-specialisation of the contemporary engineer, having evolved from the visionary master builder of the past to the number-crunching designer of individual components of today. The great structural engineers of the late 19th and early 20th centuries, such as Isambard Kingdom Brunel, Gustave Eiffel and Robert Maillart, designed holistically to invent new technological possibilities. The vision of the pure engineer as lone genius, achieving beauty through the constraints of economy and efficiency, has been celebrated by Sigfried Giedion,[2] Le Corbusier,[3] David Billington[4] and many others over the past century. The structural engineer as singular artist applies most clearly to bridge design, where the challenge of spanning allows structure to dominate the design process. On the other hand, building design requires a level of synthesis among disciplines which does not often allow structure to emerge as the primary consideration, and it is therefore more difficult to identify examples of the heroic engineer in the design of buildings.

The profession of structural engineering is in a state of open crisis today. *A Vision for the Future of Structural Engineering*,[5] published by the Structural Engineering Institute, identifies severe problems and characterises the field as occupying a 'shrinking space'. It also highlights the challenges in structures education and laments that most undergraduate curricula have not changed in decades. Compared with the staggering pace of change in computing, biomedical engineering or nanotechnology, the field of structural engineering can seem frozen in time. So it is a challenging time for structural engineering. Engineers are asked to do more with less: to deliver more design options with lower costs and lower environmental impact. And to have fewer people design more complex projects in less time. Yet, within this landscape of crisis, there are numerous examples today of stellar structural engineers bringing value to design teams.

In characterising the interwoven roles of the architect and engineer, Le Corbusier defined this as a struggle between the 'spiritual' and the 'economical' (Figure 1). Design is an endless frontier. It requires finding a balance between the pragmatic and the sublime. Architectural education emphasises the plurality of solutions and encourages exploration. Engineering education emphasises unique solutions, which can lead to a reluctance to explore. But the greatest engineers are ceaseless explorers. Today, increased computational power is allowing engineers to shorten feedback loops in design by articulating a common language for design goals and by providing a clearer view of the terrain to be explored. Instead of providing a unique solution for the design

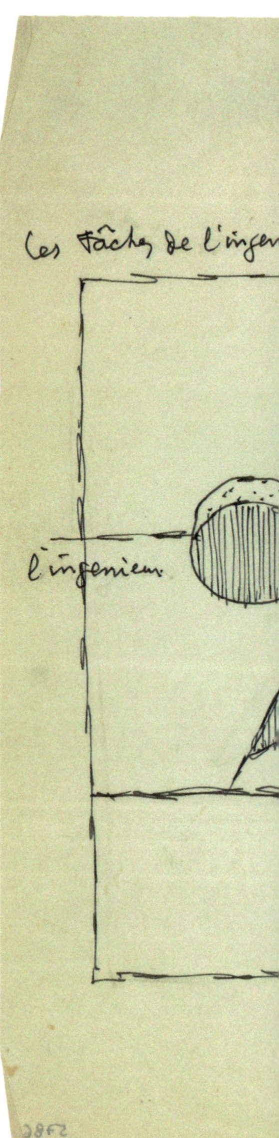

team to accept or reject, the best engineers can map the design constraints in a productive way. The exploration of the engineer is bounded by ethics: by protecting human life in building safely; by pursuing design efficiency in a resource-constrained world; and by seeking economical solutions for clients within a finite budget. Without constraint, there is no design.

This is an optimistic book. It portrays a highly creative practice exploring new frontiers in structural engineering and it provokes questions on the multidimensional roles of engineering in contemporary architecture and art. Structure is not the only driver in architecture, nor should it be. But the projects and methods described here demonstrate the myriad ways in which the mature field of structural engineering can still contribute in new ways. The book demonstrates the powerful opportunities for engineers to serve as collaborative synthesisers in the endless frontier of design. The fearless exploration of AKT II exemplifies the burgeoning potential for the structural engineer in the 21st century. Brunel would be impressed.

John Ochsendorf is Professor of Civil and Environmental Engineering at the Massachusetts Institute of Technology. He became a MacArthur Fellow in 2008.

REFERENCES
1 Charles Correa, 'In Search of Brunel', *A Place in the Shade: The New Landscape and Other Essays*, Penguin Books (Delhi), 2010, pp144–7.
2 Sigfried Giedion, *Space, Time, and Architecture: The Growth of a New Tradition*, Harvard University Press (Cambridge, MA), 1941.
3 Le Corbusier, *Vers une architecture*, Éditions Crès, Collection de 'L'Esprit Nouveau' (Paris), 1923.
4 David P Billington, *The Tower and the Bridge: The New Art of Structural Engineering*, Basic Books (New York), 1983.
5 *A Vision for the Future of Structural Engineering and Structural Engineers: A Case for Change*, ASCE: Structural Engineering Institute, 2013, http://www.asce.org/uploadedFiles/visionforthefuture.pdf

IMAGE
Figure 1© FLC/DACS, 2016

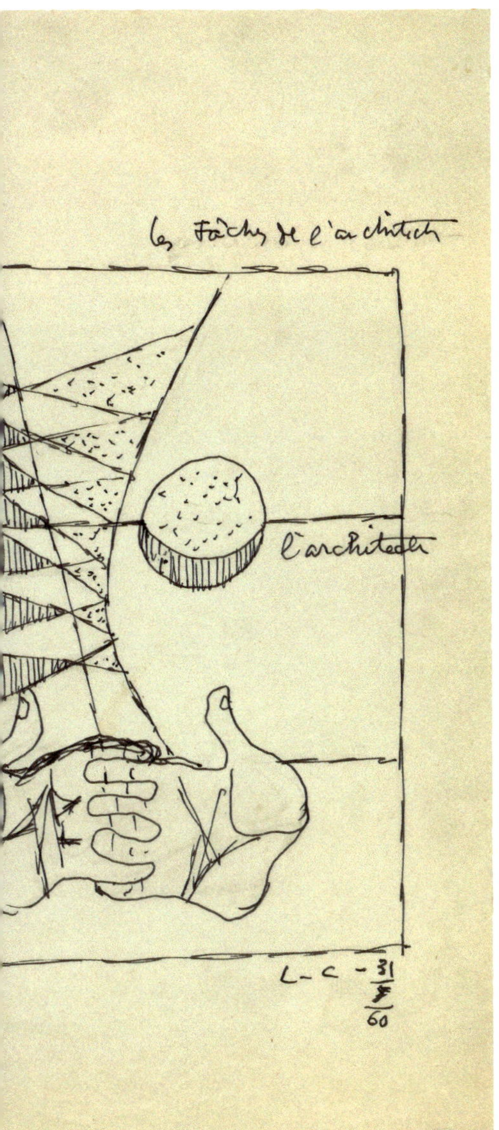

1 The yin and yang of the architect and engineer, Le Corbusier, 'Le Nouvel Aujourd'hui: les tâches de l'ingénieur et de l'architecte', 31 May 1960.

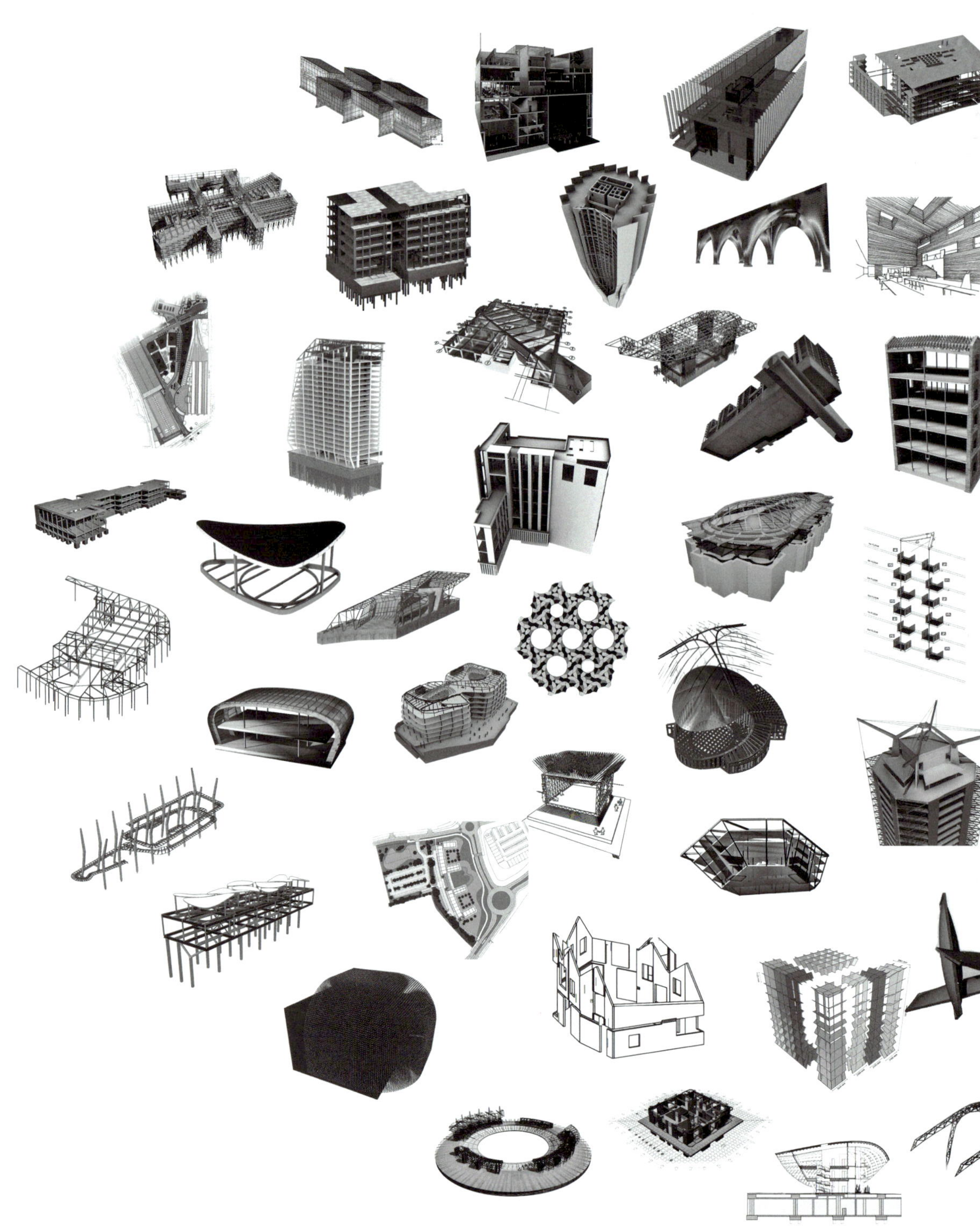

PART 1: INTRODUCTION AND TERRAIN

HANIF KARA

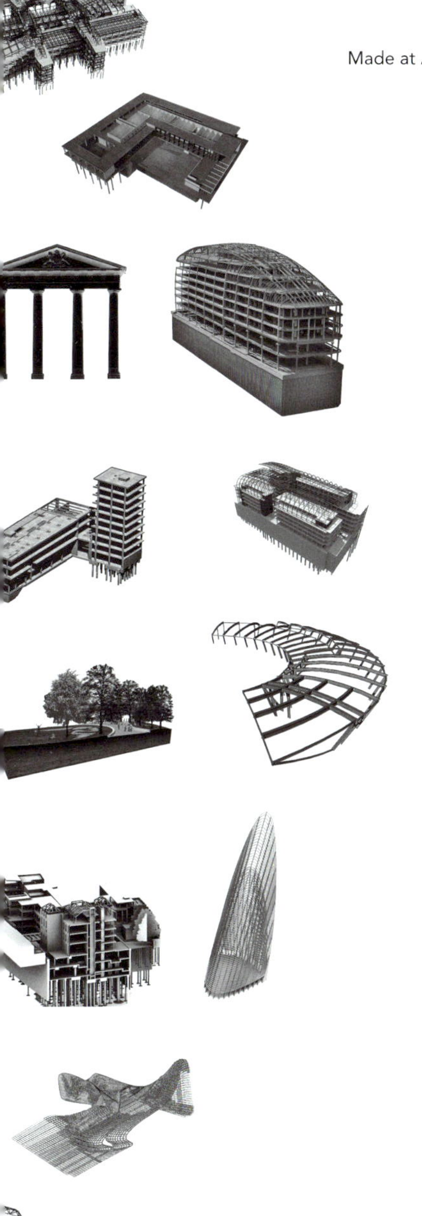

Made at AKT II

Design engineering has become a cliché of seismic proportions with multifarious and slippery meanings. In order to reassert a coherent promise and to avoid getting caught in its propagation as a buzzword, this book establishes a precise meaning from the personal viewpoint of the editors based on the fundamental triumphs, experiences, methods and concepts developed at AKT II, a leading design-led, structural engineering practice. This is executed by dividing the book into two parts, 'Introduction and Terrain' and 'Heft, Ontology and Horizon' of design engineering, defined as an esoteric scientific discipline combined with visual stimuli. To make sense of the changes in design engineering and identify patterns, we have curated contributions from past and present colleagues over two decades, enabled by a combination of practice, design research and academic encounters which capture new technologies, analytical tools and processes that have emerged. We are careful not to be conclusive about the subject.

The first part of the book sets the pace by peeking at the recent past, but focusing on the present to paint a picture of a complex blend of high-tech, low-tech, old and new, digital and analogue, with sometimes contradictory outcomes that are the terrain of design engineering today.

1 THE 'PINK NOISE' OF DESIGN ENGINEERING   HANIF KARA

The boundary and border of architecture and structural engineering have traditionally been defined by a linear and hierarchical correspondence between the two disciplines. Professionalisation of both disciplines has created a pre-articulated routinisation of the practices and distinct processes where the architect develops insights in design, while the structural engineer is granted exclusivity to react only once the design is developed. Today both are required to develop new skills and competences if they are to survive. In response to cultural and technological developments in the last twenty years, this relationship has evolved significantly, changing economic orders (where rising wealth has increased the importance of aesthetics) and, more recently, presenting new opportunities to question 'planned obsolescence' of buildings through the reshaping of design disciplines.

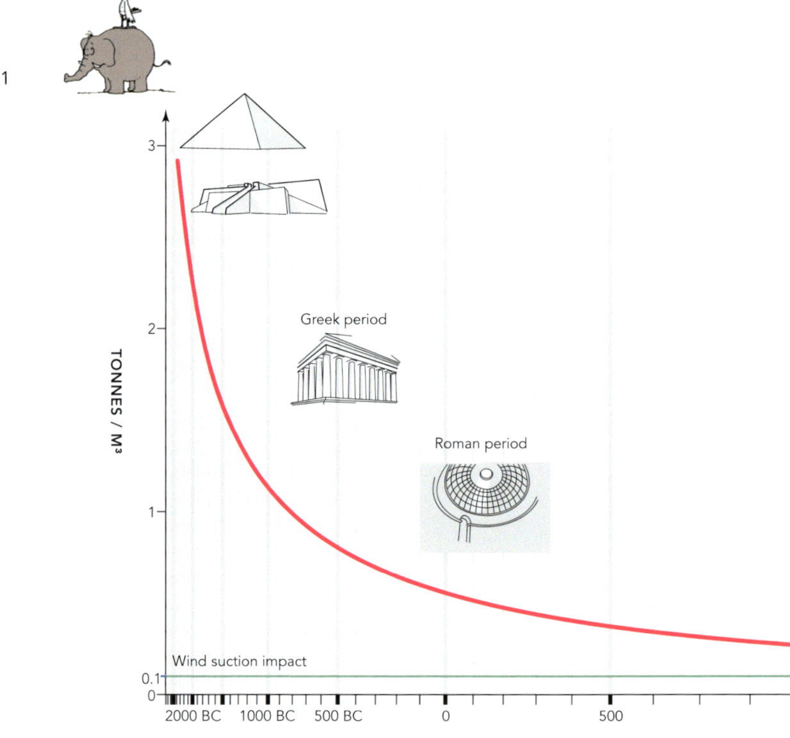

1 AKT II, birth of the structural engineer (c 1800).
Birth of the structural engineer and their position in the wider spectrum of architectural design discipline and history (which transcends our discipline).

The complex, changing relationship between the two disciplines
cannot be explained easily, and any historical appraisal of the shifts
could start in many places; we must therefore select the starting
point of such a narrative carefully. Over the last century, the most
compelling spark to questions concerning the dichotomy between
'architect' and 'engineer' as designers came from Le Corbusier
in 1927, when many believe he asserted that the process of
engineering should drive the development of architecture:

> 'Engineers make architecture, since they use calculations
> that issue from the laws of nature, and their works
> make us feel HARMONY. So there is an aesthetic of the
> engineer, because when doing calculations, it is necessary
> to qualify certain terms of the equation, and what
> intervenes is taste. Now when one does calculations, one
> is in a pure state of mind and, in that state of mind, taste
> follows reliable paths.'[1]

While one cannot agree with all of the implications inherent in
Le Corbusier's prescient statement, it was a strong encapsulation
of the prevailing feelings of the time, and we can clearly trace
their trajectory and legacy through the Modern and Post-Modern
architectural movements, as exemplified by influential figures
such as Ludwig Mies van der Rohe, Louis Kahn, Tadao Ando and
Team 4 (Norman Foster, Richard Rogers, Su Brumwell and Wendy
Cheesman), who each pushed for greater parity and collaboration
between the two disciplines.

In subsequent decades, this cultural realignment was reinforced
by profound changes in the field of structural analysis. Despite

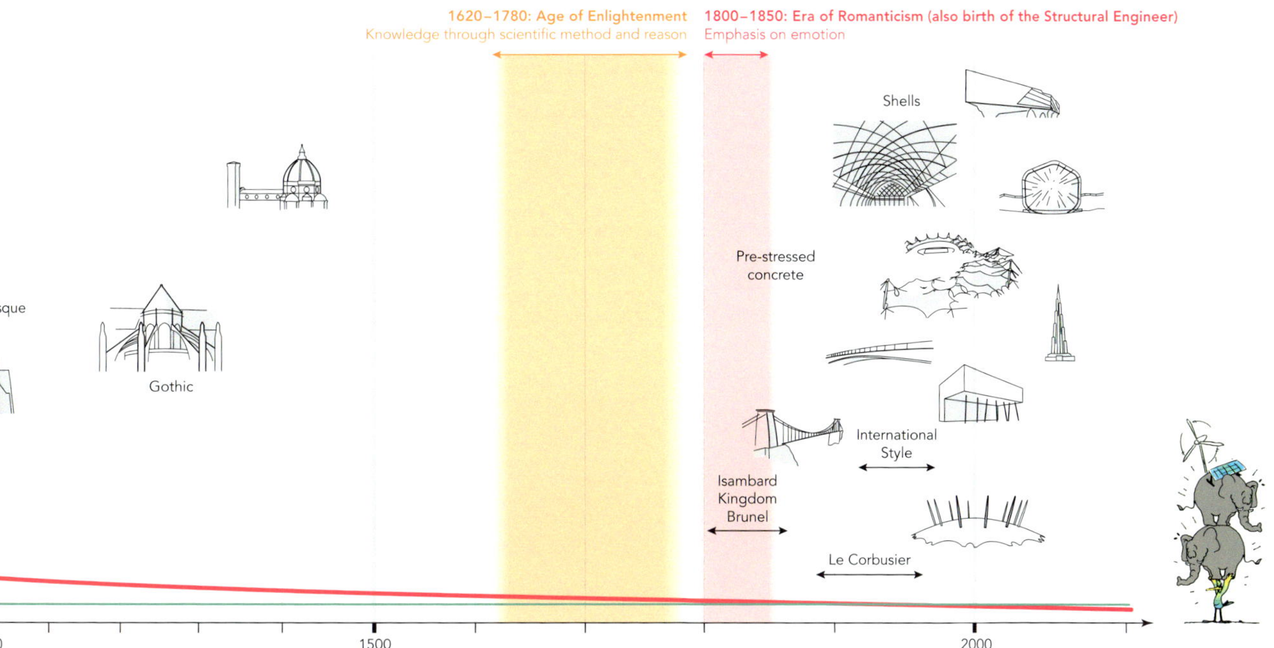

2 AKT II, evolution of reinforced-concrete
design codes of practice.
These codes have evolved since 1930, they
assess how factors of safety, combined with new
analysis, allow the reuse of old structures.

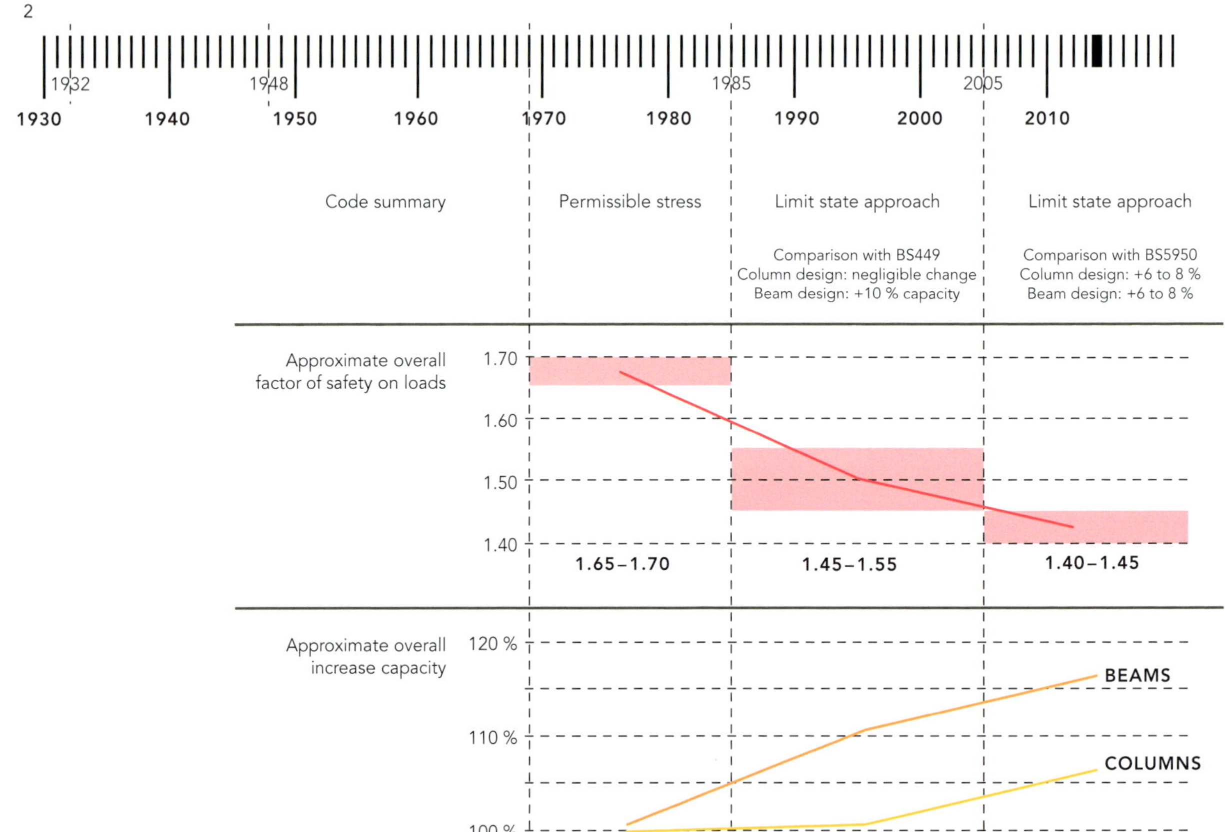

the crippling effects of the Second World War slowing progress in many areas of construction, a significant turning point occurred with the birth of the first threads of 'limit state methods'. Driven by the greatly reduced availability of materials post-war, early experimental work in structural engineering conclusively demonstrated that analysis of stresses computed with simple elastic theory was far removed from how structures behaved, and from this emerged the concept of 'plastic limit states' that would erode and, in some instances, decompose factors of safety.

Though the gulf between the design of steelwork and reinforced-concrete structures remained, limit state methods advanced both materials. Notably in the UK between 1936 and 1948, the engineer John F Baker developed plastic theory for the design of steelwork, indeed it was used for the design of Morrison shelters during the war. While these methods are no longer used for simple structures today, the principles can be used for any buildings.

Meanwhile, in 1955 in the USSR, Professor NS Streleski developed methods of limit state design[2] that led to the introduction of the first codes in reinforced-concrete design using ultimate limit state method to reduce safety factors in concrete states (Figure 2). And, though not widely used until 1965, this has led to economical structural designs.

Such advances inspired new confidence that can be seen in the work of a long line of engineers since then; people such as Ove Arup, Ted Happold, Felix Samuely, Tony Hunt and Fazlur Khan, Cecil Balmond, Mike Schlaich, Jürg Conzett, Klaus Bollinger, William Baker and Peter Rice, all went on to broaden this incipient trend in the hope of spreading the value of a 'creative collaboration'. Most significantly, Rice famously urged engineers to 'imagine' and temper the use of pragmatism to escape the characterisation of engineers as 'Iagos'.[3]

THE PRESENT CONDITION AND REDEFINITION OF DESIGN ENGINEERING
Today's fertile atmosphere provides a novel condition for the specific relationship between architects and structural engineers, as both disciplines try, breathlessly, to keep up with the pace of change. During the early 1990s, newly awakened powers of observation and increased skill in representation encouraged both disciplines to 'peek' into each other's work again in search of perennial reinvention. At the height of this period, the boundary between uniquely human creativity and machines' capacity for pattern-recognition and complex communication marked a new confidence, offering freer movement between the two disciplines, and between design fabrication and construction. As both platforms and protagonists, leading structural engineering design offices, design schools and educators play a big part in this dance of the disciplines. What was noticeable in the first 'wave' (1990–2000) is that architects, in response to the popular imaginations of their consumers, were increasingly expected to exemplify with each project a 'newness', 'cheapness', 'particularity' or 'uniqueness' for the production of 'one-off' creations (often

formerly unimaginable forms) that avoided universality. Meanwhile, other abundant productions of architecture, such as housing in emerging markets, continued – due to rapid urbanisation – with very little design and often without architects.

At the height of this trend, in his controversial thesis of 2002, Stephen Wolfram even stretched the traditional approach of computation, through mathematics and engineering, to empirically investigate computation for its own sake. Though seen by many as an 'abrasive approach', it did give valuable insights and observations: 'Whenever one sees behaviour that is not obviously simple – in essentially any system – it can be thought of as corresponding to a computation of equivalent sophistication.'[4]

The opportunities for structural engineers and technologists to support the endeavours of architects expanded in response. It is clear that initially, to a greater or lesser degree, even structural engineers were guilty of being stuck in a tectonic discourse, often using the same technologies to produce inanimate aesthetics driven by the latest software, prestige and abundance of resources, sometimes fuelled (in part) by undiscerning constructions in developing economies.

Simultaneously, this expanded opportunity allowed some engineers to grow their own disciplines freely, encouraged by the extraordinary freedom to ransack the software chest in search of the thinnest glass, shallowest curve, longest span, and so on. While such expansion has to be tolerated, in many cases it resulted in architects and structural engineers working in an atmosphere of unclear thought and sensory profusion, encouraging the self-sabotage, gimmickry and posturing of so-called 'archineers' and 'engitects' (Figure 3). At the start of any new-found freedom is a 'big bang' effect, setting free a certain amount of pent-up demand. The Beijing Olympic Stadium is an example of this; looking back at this new structural wonder, one has to question its provocative deception in the use of steelwork – a 60 mm x 2 mm strip that could wrap around the globe three times. In hindsight, we believe this approach failed to engage with the larger, more fundamental behavioural changes on offer to us as designers.

3 AKT II, prism.
In present-day practice, technology and specialisation have proliferated the process of design to an extremely 'thin slicing' of architecture. We hark back to previous traditions in the work shown here.

3

THE DESIGN ENGINEER AS PRACTITIONER
To be most productive in this new paradigm, we chose to synthesise the making of buildings at one extreme, and engagement in the discourse of design at the other, allowing us to practise our own 'behavioural design engineering' that incorporates aesthetic, linguistic and technological spaces within practice. This approach takes on a more comprehensive and universalist interpretation of design than that circumscribed by normative disciplinary behaviour of the structural engineer, while keeping in mind preceding successes and failures. Building a practice that can tune in to this behaviour has encouraged significant creative achievements.

On the subject of aesthetics, for instance, we saw clearly that the traditionalism of our discipline had led it to be perceived as polarising, as too reliant on finite technologies that give binary answers, and overly decisive in practice. The first challenge for us has been to banish these perceptions, and to increase acceptance that design, from an engineering perspective, is as much a visual subject as a scientific one. It requires a long and deep immersion in our own discipline, but also an appreciation of when that discipline ceases to be appropriate in the creation of the best buildings. This requires both engineers and architects to engage in a forum of interaction, thought development, research and qualitative outreach, while avoiding 'switching disciplines' or 'crossing over'.

This particular stance allows us to re-engage with the architect and guards against premature optimisation by promoting engineering as a less than exact discipline, and by utilising a more unconstrained approach in which solutions are not determined by calculations alone. Such shifts engender a longer and more rewarding conversation between architect and engineer in which both parties are able to circulate and refine design options, freed from the spectre of the looming 'design freeze', representing a greater value to our immediate client.

In this endeavour we continue to be guided by the foundational engineering theories (the equilibrium of internal and external forces, a clear understanding of geometry and boundary conditions, and the knowledge of material properties). These are now supported by a more refined appreciation of the other disciplines involved, which fosters an interdisciplinary atmosphere able to take on the broadest agenda of design innovation. Out of these changes we believe a new kind of engineer emerges: the 'design engineer'. Able to see a project in the architect's terms, but with the mind and eyes of the structural engineer, they produce holistic solutions that integrate all aspects, rather than residing in a particular system, element or tool. The design engineer's scope is no longer limited solely to the manipulation of building materials and processes, but incorporates technology, skills and knowledge of the dialectic relationship between nature and the superimposed built environment.

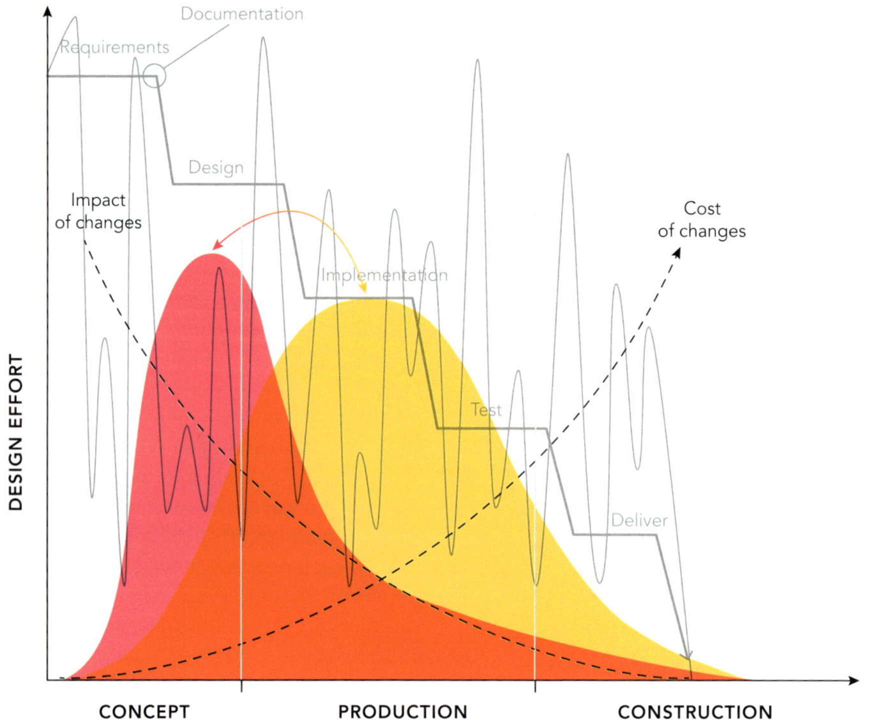
Documentation
Requirements
Design
Impact
of changes
Cost
of changes
Implementation
DESIGN EFFORT
Test
Deliver
CONCEPT
PRODUCTION
CONSTRUCTION
CLIENT CASH FLOW 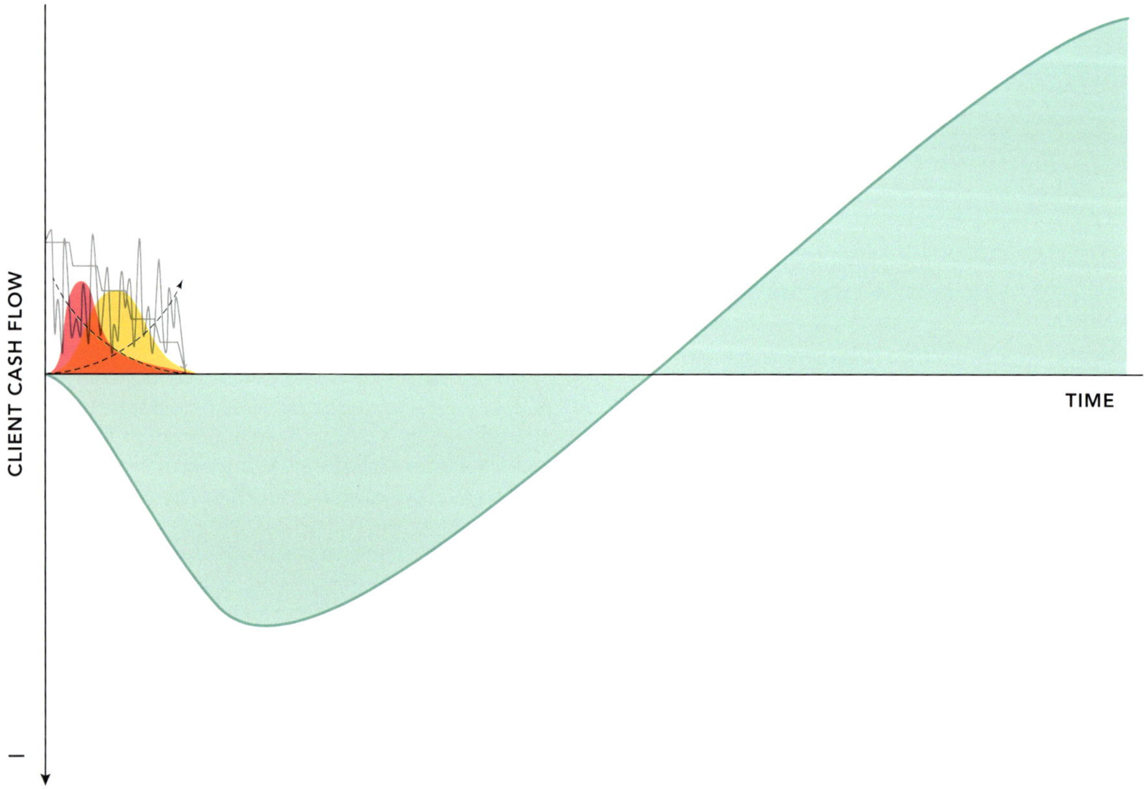
TIME

Alongside these practical aspects, we also wrestle with the pedagogical implications. It is apparent that the current model for the formative education of structural engineers does not go far enough to develop design engineers, as it is based predominantly on scientific knowledge. It should be reformed around the 'creative design process', retaining ancient wisdoms while embracing new opportunities such as digital design and manufacturing techniques twinned with the brute force of computation, and a deeper understanding of natural and high performance materials through physical and digital testing. Yet this only scratches the surface of the possible implications: what role do codes of practice play? How does this affect the practice of structural design organisationally? Is the discipline sustainable?

PARS PRO TOTO

From the formation of the practice in 1996, we recognised that it was hard to turn insights in designs into engineering without dissecting existing models of practice and understanding the design value chain (Figure 4). This required us to combine a staunch 'for profit' endeavour with a 'non-profit' behaviour that would allow the one to fund the other. Within this model, one of our key responses to these disciplinary changes was to create a (not for profit) network inside the practice that we have called 'p.art' (applied research team) (Figure 5). As its name suggests, p.art was conceived as an integral element within the framework of AKT II; a flexible, multidisciplinary grouping with the abilities necessary to seek out and solve emerging design challenges,

4 AKT II, value of design and non-linear processes.
Diagram extended to show relationship between design and client's return value.

5

5 AKT II, p.art's role.
AKT II p.art's many activities funnel into a version of 'design engineering', taking on a new mandate of design practice greater than conventional disciplinary behaviour of structural engineers.

and to pass on those solutions to the practice as a whole. P.art fosters a discipline where we can no longer be merely mechanical or methodical, but have to bring to bear whatever skills will be needed in design, project by project. It no longer depends on conventional behaviour, but has to accommodate unarticulated desires and unnoticed influences on architecture (Figure 6).

In operational terms, the core of p.art is multi-faceted, comprising design engineers and architects, parametric designers and software developers, mathematicians and geometricians, graphic designers and writers. Other members, drawn from the wider pool of engineers, technicians and designers within AKT II, cycle in and out of the core team on a project-by-project basis, teaching and being taught new processes and techniques which they then disseminate within the rest of the company and beyond. In this way, the entire structure and remit of p.art avoids the 'siloing' of information seen in some of the early engineering industry forays into computational design, and ensures that they act as a catalyst for investigations between teams and companies, between academia and practice, and between design and construction.

While the size and composition of p.art ebbs and flows – responding to both the short-term requirements of individual projects and the larger cycles of industrial change – we maintain at all times a constant presence in academia, both through

6 AKT II, design research: scientific vs architectural endeavour (adapted from William Caudill).
We operate between these boundaries, keenly trying to combine cultural mystiques, imagery, science and new possibilities.

7 Foster + Partners with AKT II, Masdar Institute, Abu Dhabi, UAE, 2010.
The design and construction of the facade required a high-quality finish and complex computation.

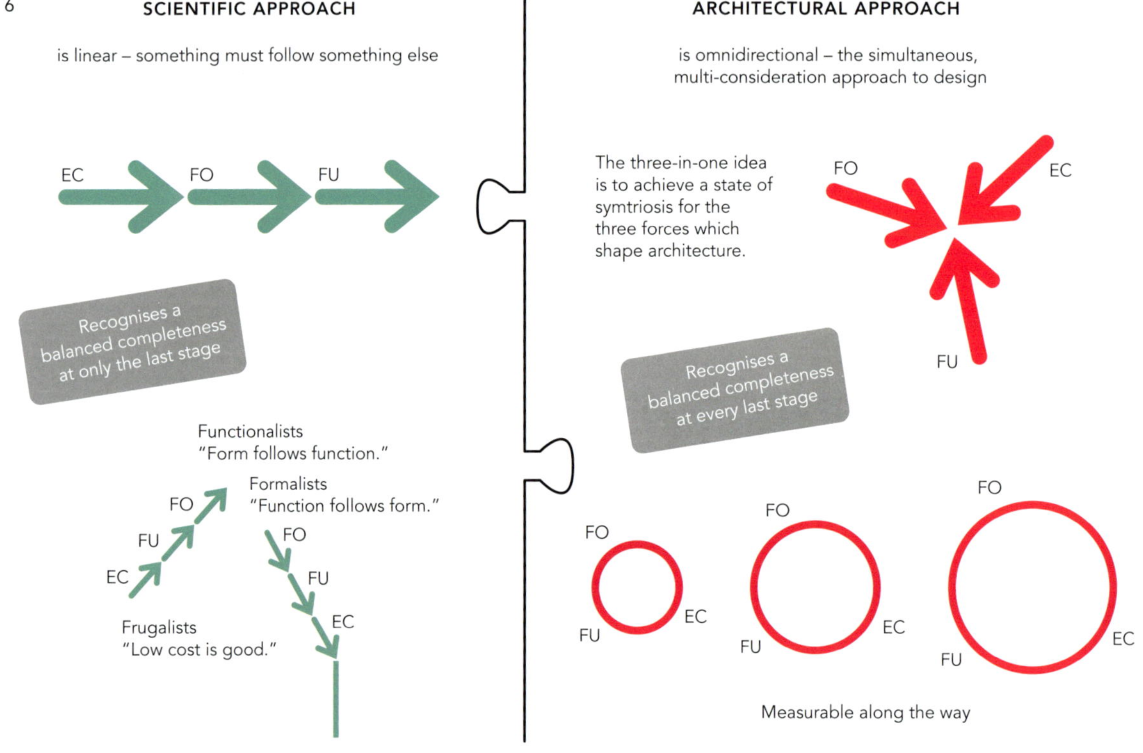

1 THE 'PINK NOISE' OF DESIGN ENGINEERING 22–23

John Lewis
John Lewis
WELCOME
TO GRAND CENTRAL

8 AZPML with AKT II, Birmingham New Street station, Birmingham, UK, 2015. A geometry that designs a 'form' of complex geometry, extending the use of advances in software and the connections between digital manufacturing and design.

9 Asif Khan with AKT II, Radiant Lines
installation, Melbourne's Federation Square,
2014.
With Asif Khan, work such as this always
requires the engineer to peek into other senses,
such as light (as in this case) or sound.

9

tutoring at a number of international design schools, and through selected partnership with material laboratories, postgraduate research, and so on.

The 'embedded' position of p.art within the practice has been critical in redefining what makes a design engineer in particular. By helping to clarify intent without relying on science and calculus alone, their expertise in specialist areas has helped us to deliver projects that bring to the foreground the role and value of design in engineering through interdisciplinary interaction, bespoke non-linear processes and the expansion of transitional convention between structural engineering and other design disciplines today. One of p.art's other roles has been to push the theoretical accuracy of calculation beyond certain limits in a proportionate manner that makes it useful on a project-by-project basis. Recognising the distinction between 'basic' knowledge and 'interdisciplinary' knowledge, this behaviour brings focused design engineering to each project, in different degrees, to act as the bridge-builder between disciplines.

SUSTAINING THE OUTCOMES THROUGH ECONOMIC CHALLENGES TODAY
Out of this invigorating and occasionally tumultuous history the selection of work documented in this publication has emerged: a wide-ranging, yet coherent, set of projects, each expressing a desire to escape ugliness at many levels and embrace those differences that ensure a productive relationship of greater creativity and utility, rather than of obviousness.

We have welcomed the opportunity to let outside voices – as well as members of p.art (past and present) and the wider organisation within – contribute to this discourse. In addition, and to enable a demonstrable outcome, we have taken a position that, whenever possible, stitches 'scientific research' with 'design research', given that scientists and engineers are largely dismissive of the latter.

The economic crisis of 2009 forced us to look back at the previous decade to find evidence of the value that has come from this recent development of our own practice. This was needed in order to continue building on the intelligence gathered to date and to discover whether the didactic air of the digital era cloaked some exuberance that we needed to remove. On one hand, the tools had been used as party tricks (colourful analysis dressed as design) for promoting design to a commercial level but removed from the source of good engineering. But on the other hand, as demonstrated by the Heydar Aliyev Centre project in Baku, the Masdar Institute (Figure 7) and Birmingham New Street station (Figure 8), bespoke tools have allowed us to deliver remarkable architectural visions, even in the most difficult and remote environments. The Radiant Lines project (Figure 9), the Bivak in Slovenia (Figure 10) and Hunsett Mill in Norfolk (Figure 11), are all projects that dissolved the boundaries between screen and workbench in their production and redefined approaches that deal with extreme environments with combinations of high- and low-tech experimentation. At another technical extreme, the Angel Building

10 OFIS Architects with AKT II, bivouac, Slovenia, 2014.
GSD Harvard students: Myrna Ayoub, Oliver Bucklin, Zheng Cui, Frederick Kim, Katie MacDonald, Lauren McClellan, Michael Meo, Erin Pellegrino, Nadia Perlepe, Elizabeth Pipal, Tianhang Ren, Xin Su, Elizabeth Wu. Originally conceived to reach the site by 'drones', eventual construction on site used a conventional helicopter.

11 ACME with AKT II, Hunsett Mill, Stalham, UK, 2010, rear elevation.
Difficult site access exploited the use of easy to assemble flat-pack engineered timber, next to a traditional protected masonry building.

12 Allford Hall Monaghan
Morris with AKT II, BAM,
Angel Building, London, UK,
2010.
An existing building that was
transformed to set a new
benchmark in what is possible
by connecting complex
analysis, new materials and an
understanding of reinforced
concrete.

in London used changes in codes, combining this with advanced composites to breathe new life into an old building (Figure 12).

In the case of the Heydar Aliyev project (Figures 13 to 15), we designed the frame with the most advanced analysis, but, in what could only be described as the 'height of sophistry', the final construction involved the use of a space frame, claiming to be more economical. Sometimes designers have to accept that a construction is not what is designed. On the face of it, Turner Contemporary (Figure 16) in the United Kingdom is perceived to require little structural engineering, but in contrast to Heydar Aliyev where the effort is explicit, many options had to be found to get to this apparently effortless conclusion using the same methods and tools.

Structural engineering retains an intrinsic value through remaining non-restrictive, mainstream and confident at its roots. In a world where design is now everywhere and everything is designed, the generalisation of the term 'design engineer', and the characterisation of what it implies in practice, has recently blurred the boundaries to a point where it has become 'general background noise' that is self-defeating, and homogenises what engineers can contribute. By selecting a narrower approach in this publication, we want to refocus design engineering onto what we call the 'pink noise' to distinguish it in the field, not claiming a status, but verifying a status claim (Figures 17 and 18). The fundamental premise of that status claim is born out of a behavioural change in our practice which acknowledges that

13

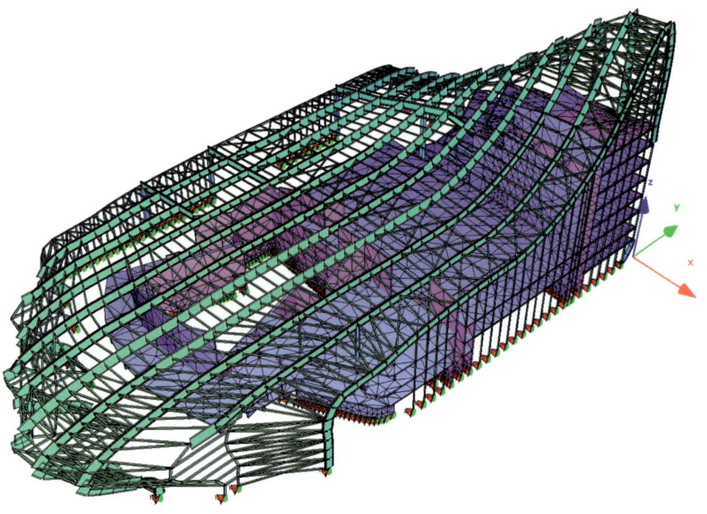

14

13 Zaha Hadid Architects with AKT II, Heydar Aliyev Centre, Baku, Azerbaijan, 2012.
The completed building required advanced scripting methods and analysis for sensitivity in developing countries.

14 Zaha Hadid Architects with AKT II, Heydar Aliyev Centre, Baku, Azerbaijan, 2012, Heydar Aliyev Centre analysis.
The design was based on simplified linear frames to cope with the complexity of the facade. Facade tiles were optimised to maintain overall form but economy in manufacture and installation.

15 Zaha Hadid Architects with AKT II, Heydar Aliyev Centre, Baku, Azerbaijan, 2012.
Final construction introduced off-the-shelf space to construct a unique form, challenging the purpose of space frames.

16 David Chipperfield Architects with AKT II,
Turner Contemporary, Margate, UK, 2011.
Annual overtopping of the water required
careful modelling for extreme conditions.

architecture transcends engineering and can easily ignore, avoid or escape the space occupied by design engineering; but on the other hand, our particular version of design engineering is required to navigate the space occupied by architecture as a condition of its existence. We have to be cognisant of its success and failures, and resist the temptation of crossing into the realm of architecture in order to reinforce its position in 'specialised interdisciplinary' discourse. To add value from this position, based on the evidence of the completed projects over the last twenty years and on the board at present, we hope to trigger a small change in the education and practice of design engineers as we know it today, taking it beyond institutionalisation and professionalisation, and away from being another annoying trend.

REFERENCES

1 Le Corbusier, *Toward an Architecture*, J Paul Getty Trust: Frances Lincoln Edition (London), 2008.
2 Limit state design (LSD) refers to a design method used in structural engineering. A limit state is a condition of a structure beyond which it no longer fulfils the relevant design criteria. The condition may refer to a degree of loading or other actions on the structure, while the criteria refer to structural integrity, fitness for use, durability or other design requirements. A structure designed by LSD is proportioned to sustain all actions likely to occur during its design life, and to remain fit for use, with an appropriate level of reliability for each limit state. Building codes based on LSD implicitly define the appropriate levels of reliability by their prescriptions.
3 Peter Rice, *An Engineer Imagines*, Ellipsis London (London), 1996.
4 Stephen Wolfram, *A New Kind of Science*, Wolfram Media (Champaign, IL), 2002.
5 https://en.wikipedia.org/wiki/pink_noise [accessed 4 April 2016].

TEXT

IMAGES

17

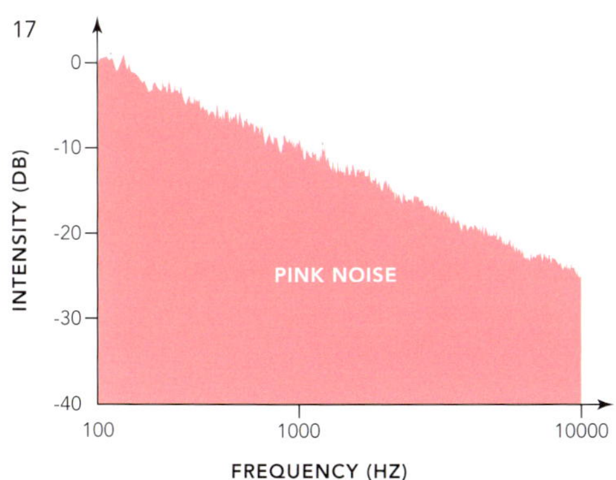

18

17 AKT II, pink noise graph.
Pink noise can mask low-frequency background sound, helping to increase one's productivity and concentration. The themes and projects here are intended to mask the wide-ranging disciplinary activity of structural engineering.[5]

18 AKT II, pink noise function.
We borrow the use of this 'function' as a metaphor to make a distinction between the general implications of 'design engineering' and the specific approach discussed in this publication.

2 DIGITAL TO POST-DIGITAL DANIEL BOSIA

From the work of Pier Luigi Nervi, Buckminster Fuller and Frei Otto (Figure 1), we have learned how research into natural systems with a 'scientific' approach can generate new and unexpected expressions in architecture, promoting a deeper dialogue between the worlds of art and science. The work of these three pioneers remained rather tangential to the Modernist and Post-Modernist movements in architecture, whose focus was the development of stylistic and philosophical responses to the uncertainties of the first post-war period. It was through the work of engineers such as Ove Arup that the discipline of engineering started to influence architecture on a global scale. Arup understood the importance of 'total architecture', the integration of both disciplines.

1

2

1 Frei Otto, Mannheim Multihalle, Mannheim, Germany, 1975.
Gridshells are lightweight single-layered compression structures. Mannheim Multihalle was constructed by using layered straight timber joists, bent by lifting of the canopy on site and allowing it to shear at the joints. The structure was then locked at the joints with simple bolted connections and restrained at the perimeter with a tension ring.

2 Grimshaw, Eden Project, Cornwall, UK, 2001.
Space structures are double-layer structures consisting of an outer compression layer and an internal tension layer, spaced apart by diagonals. In the Eden Project, the inner layer was realised with tension rods to minimise visual impact. The cladding is realised with long-span lightweight ETFE pillows.

Today, we know that multidisciplinary design is a much more fluid and dynamic concept than that of the large multidisciplinary 'one-stop-shop', and relies on the ability of leaders in their sector to overlap and intersect with their peers in other disciplines. With projects like the Sydney Opera House, the rational engineered form has become the iconic expression of a new aesthetic, and the Centre Pompidou and the Eden Project (Figure 2) have given rise to a whole new high-tech engineering style; through projects like these, Peter Rice, Ted Happold and Tony Hunt were able to transform engineering into art and poetic invention. Their ability to turn accepted ideas on their head, and their rigorous mathematical and philosophical logic, made them some of the most sought-after engineers of our time. Others, such as Cecil Balmond, extended this model through high-profile collaborations with influential architects and theorists such as Rem Koolhaas and Toyo Ito (Figures 3 and 4).

At the turn of the century, groups were formed in architectural and engineering firms to bridge the widening gap between these two traditional mainstream disciplines in a fast-evolving technological scene. These groups embraced new computational methods to create rigorous geometric architectural designs, while also experimenting with early examples of digital fabrication. From parametric modelling to generative design and more sophisticated processes of form-finding and topological optimisation, research-based design groups paved the way for a more widespread use of computational design, developing shared languages and protocols between architects and engineers.

3

Wall A
Wall B
Roof Plan
Wall D
Wall C
1A
4A
5A
4D
3A
6C
6A
7
5B
7
5D
2A
7
6B
4B
6B
4C
5C
3B
1B
2B

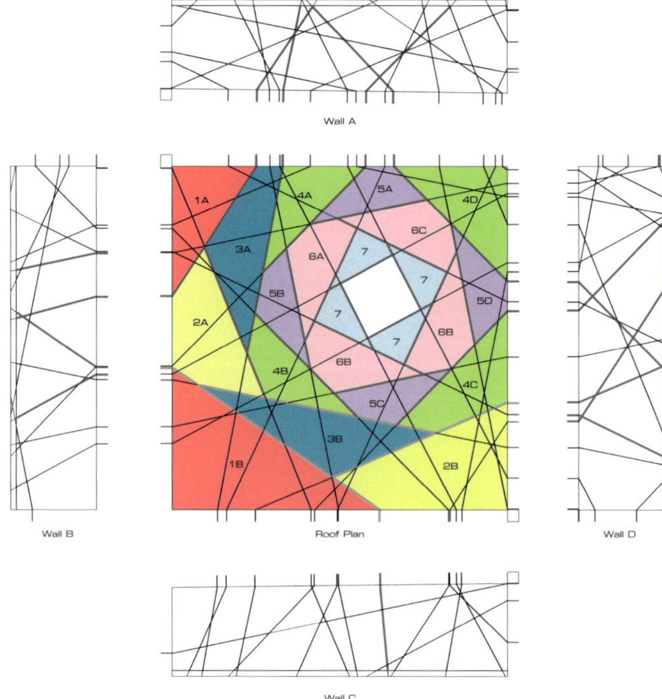

4

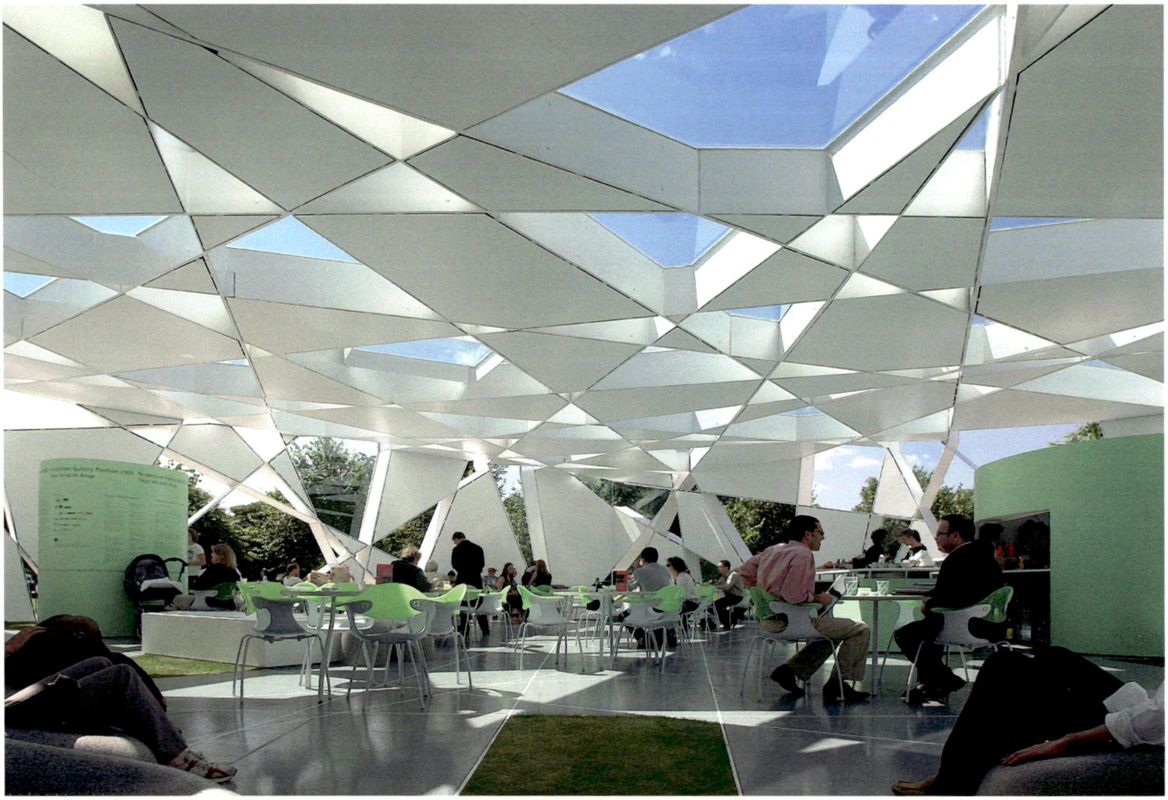

While early geometric formal explorations helped to expand the vocabulary of contemporary architecture, it took longer for rigorous digital analytical methods to be incorporated into process-driven engineering design workflows. Continually in search of new aesthetic expressions, architects keenly appropriated new digital tools, creating a flurry of innovative architectural forms, while the more systematic, and sometimes conservative, engineers preferred the more tried-and-tested analytical methods. Through the development of bespoke interfaces between the architectural modelling programs, such as Rhino/Grasshopper and the analytical software packages, interactive feedback loops could be formed between geometric modelling and analysis in a truly integrated design process.

Since the year 2000, there has been an explosion of 'digital design' in academic explorations and prototypical physical manifestations, from small pavilions to larger installations. There are few examples where digital design has been used successfully, economically and sustainably in architecture, addressing the well-being of its human users and enhancing the surrounding environment. At worst, digitally designed buildings have become physical manifestations of the tools that created them; at best they have started to demonstrate the kind of new forms of spaces and architectural conditions these can create, questioning the classical and Modernist generic models. In the Arnhem Centraal station by UNStudio, the roof of the main transfer hall peels down to form a structural support in the form of a mathematical Seifert[1] surface, acting as a fluid channel of structural forces, people, light and air (Figures 5 and 6).

We have become accustomed to parametric fields of geometrically morphing cladding components across 'organic'-looking forms, doubly curved NURBS[2]-like surfaces or geometric tiling patterns that have been, and will continue to be, important iconic examples of a new digital era. However, with the heritage of the last 15 years, we have now confidently entered a post-digital era, in which digital design and fabrication methods are part of our daily lives and working methods; they no longer need deliberately to display the digital methods with which they were modelled on the computer or fabricated in the factory. They don't need to be celebrated as 'special' or 'new' any more, because those methods

5

5 UNStudio, Arnhem
Centraal station, Arnhem, the
Netherlands, 1998.
Catmull–Clark algorithm of
the geometry. Borrowed
from computer graphics, this
mathematical subdivision
algorithm was used to
generate the smooth surfaces
and seam patterns of the
station roof. The algorithm
was adapted to allow control
of the geometry at the
perimeter. The Catmull–Clark
smoothing algorithm –
adopted in Arnhem station
for the first time in a building
design application – is today
implemented by several
computer programs such as
Maya and Rhino.

6 UNStudio, Arnhem
Centraal station, Arnhem, the
Netherlands, 1998.
The central support to the roof
of the Arnhem station consists
of an extension of the surface
to the concourse level and
platform level in the form of a
mathematical Seifert surface. In
such a non-orientable surface,
the top meets the bottom in a
continuous flow.

of design are inherent in all of a new generation of buildings we are designing and constructing. It is not only 'special' buildings that benefit from these tools and processes, but all of them to a greater or lesser extent. But at the same time, the complete rejection of digital methods in the way we work and conceive of buildings is naive and counterproductive, as it leads to forms of stylistic romanticism that are often self-referential and inhibiting in today's discourse on architecture.

The intersection of the digital with the physical (through the advancement of rapid prototyping technologies and other automated fabrication methods), together with the intelligent interface of digital methods with manual 'crafted' processes, has brought the human-designer back into the driving seat. Through a process of appropriating new methods of design and fabrication that allow the hybridisation of digital and analogue techniques, designers and fabricators have been able to establish new interfaces. Similarly, the dialogue between art and science has found new meanings through the work of artists such as Olafur Eliasson, Anish Kapoor and Matthew Ritchie. In *The Morning Line*, Matthew Ritchie created a piece that, through the use of a mathematical three-dimensional tiling, attempts to represent the entire structure of the universe in its multiple embedded fractal dimensions and patterns (Figures 7 and 8). Today, the artificial and natural worlds are hybridised through the development of robotic and drone technology, transforming the way buildings are, or could be, constructed. Synthetic and organic materials are now being introduced with enhanced structural and environmental properties; these include up-cycled timber from renewable sources or super strong carbon-fibre composites to replace traditional steel and concrete materials. The sophistication of digital tools, their improved interface with human interaction and the real-time feedback that these allow, have reintroduced the 'analogue' within a digital workflow. In other words, the designer is both part of and in control of the process; he is the choreographer, whose work is enhanced by digital means. This allows him/her to make informed decisions, through access to knowledge and data. The designer or design team are masters of their digital tools, not slaves to pre-structured software packages; they are the creative 'architects' of hybrid, intuitive and rational processes to realise better and more holistic spaces for human habitation and fulfilment.

Today, thanks to a whole new generation of 'computational designers', we have the ability to design our own processes and tools, rather than accepting the inherited traditional answer which has been imposed on us by software companies. Computational design has allowed us a much closer interaction and feedback loops between analytical processes and design, to the point where they have established an intimate dialogue. This is evident in many projects, where the analytical work of the engineer is closely linked to that of the architect. In the Central Bank of Iraq in Baghdad, designed by Zaha Hadid Architects with AKT II, a complete workflow was created that allowed the real-

7 Matthew Ritchie with Aranda\Lasch, The
Morning Line, Seville, Venice, New York,
Istanbul, Vienna, Karlsruhe, 2009.
The Morning Line is an experimental project
that explores the interdisciplinary interplays
between art, architecture, mathematics,
cosmology, music and science.

7

8

time structural form-finding and geometric optimisation of the concrete petal-shaped exoskeleton, which was designed to resist blast loads at the base and open up to receive daylight at the top of the tower (Figure 9).

New interaction methods allow an embedded performance-based approach to design that addresses the complexity and specificity of modern building. These methods also open the door to a more integrated approach to flexibility and adaptability. The essential need to transform and customise one's working or living environment throughout its life is key, not only to personal well-being, but also to the building's efficiency and sustainability through its 'intelligent' interaction with, and adaptation to, its surrounding environment.

'Multi-parametric design' is now necessary to address the complexity of human needs and to face the economic and environmental challenges of modern society. Efficiencies derive from the understanding of the interaction of different, sometimes contradicting, forces. A holistic approach to architecture is necessary, since modern buildings and cities are complex organisms that need to respond to a multitude of specific local conditions. A siloed approach between disciplines is no longer possible. For this reason, each discipline has to establish deep roots and interfaces with the other so that the tension between different aspects can be resolved in the whole. 'Cross-disciplinary' design allows experts in each discipline to interface and develop mutual languages. Again, digital design has helped this process, as shared protocols have been developed to communicate design. Geometry, materiality, structure and environment can be synthesised into interoperable models that allow those exchanges.

8 Matthew Ritchie with Aranda\Lasch, The Morning Line, Seville, Venice, New York, Istanbul, Vienna, Karlsruhe, 2009. Truncated tetrahedron unit. This moving installation is based on the aggregation of a truncated tetrahedral tile at four different scales in a fractal geometric configuration.

Form finding
Optimization
Geometry
Grouping
Properties
Supports
Loading
Re.AKT II
Structural
dyn analysis
Fluid
dyn analysis
Thermo-dyn
analysis
CAD
BIM
...

9

10

9 Zaha Hadid Architects with
AKT II, Central Bank of Iraq,
Baghdad, Iraq, from 2011.
Shaped by potentially extreme
seismic and blast loads, the
tower consists of a solid
concrete shell at the base,
opening into a series of petal-
shaped columns to support
the larger floor plates, while
maximising views and daylight
at the upper levels.

10 AKT II, the AKT II Re.AKT interoperable
workflow.
The Re.AKT workflow, developed within p.art,
is based on an interoperable framework that
connects parametric design routines with
analysis and production packages. This has
allowed the implementation of form-finding
processes to be used in the design of the tower.

Re.AKT is the interface between the parametric geometric model
and the analytical engineering packages, designed to carry
information about material properties, local loading conditions
and other site-specific boundary conditions; it feeds data back
to the design model about its structural or environmental
performance, allowing informed design decisions (Figure 10).
Similarly, design and execution processes are merging so that
the handover to fabricators and contractors is smoother than with
traditional methods. Design engineers facilitate that transition
through tools like Re.AKT, incorporating the intelligence of
construction and fabrication processes within design from the
outset and, at the other end, ensuring that the information
is produced in a form that is compatible with the logic of
procurement and production.

We have developed universal and generic frameworks of
communication between disciplines, yet our approach to design
and our use of these interface protocols demands specificity, as
the rigid codified approaches that have been used by architects
and engineers for the past century are not flexible enough to
address today's complex challenges. In structural engineering,
for example, codified ultimate-limit state design has allowed
all buildings to be 'brought up to standard'; however, this has
meant that most of the structures designed and constructed to
date are redundant, sometimes even inappropriate, due to literal
applications of generic codes of practice. The introduction of
new, more sophisticated analytical tools, the interface of these
with design tools and the corroboration of digital simulations
with physical prototyping and testing, has allowed us to start
optimising the way we design and construct structures. In an
administration building in Africa, new form- and pattern-finding
methods were developed to create efficient seismic-resistant
shells, together with automated fabrication methods that could
be employed on a large scale and in a market where the use of
concrete is still relatively basic (Figure 11). In the past decade,
our approach to the design of buildings has transformed from
a fully codified approach to a forensic research-based one that
starts from first principles. We have gone back to the basic
elements that constitute our profession: geometry, organisation,
materiality, structure and environment. By re-analysing these with
new computational tools, we have made them accessible to the
designer by means of real-time interaction tools.

Efficiency has taken on new meanings in a world where human
and natural resources are increasingly scarce and precious.
As designers, we have to operate a paradigm shift in the way
we approach design in order to respond to these challenges.
Efficiency is no longer associated purely with the financial
economy of a structure, but with its environmental footprint,
and it has a deep impact on the well-being of the people that
inhabit the structure and the surrounding environment. We are no
longer looking at crude optimisation processes on preconceived
designs, but at engrained efficiencies within the very concept of
architecture that generate deep environmental and economic

11 AKT II, form- and pattern-finding.
Form- and pattern-finding methods have
been developed at AKT II that use shape-
optimisation algorithms, based on stress
analysis and topological optimisations, that
derive efficient ribbing patterns within shell
structures. This research is being applied on
different projects within the office.

12 Foster + Partners with AKT II, stair of
the Bloomberg headquarters, London, UK.
Geometrically the stair is a 50-metre
diameter trefoil knot, developed into a
spiral. Structurally the stair is a monocoque
structure, composed of prefabricated
interlocking segments, bolted together on
site in a staggered arrangement in order to
create a continuous rigid system.

12

benefits. An 'efficient structure' cannot be one that causes the
least damage to the environment or costs the least, but one that
improves and regenerates it.

Modern buildings need to be beautiful, flexible, lean, logical and
based on the efficient use of material resources. Their complexity
should be based on simple principles compounded to create an
articulated and flexible whole. For this we are actively developing
ways to embed efficiency within the language of architecture, so
that this can become a core value of our designs. While looking
at new and more efficient uses of material within architectural
form, we are developing interfaces with its environment. A first-
principles approach to fluid and thermodynamics allows us not
only to use a more refined process for the assessment of wind
and thermal loads on structures, but also to develop solutions for
the envelope that are structurally and environmentally efficient.
For example, the Bloomberg headquarters in London, designed
with Foster + Partners, uses a 'breathing facade' which supplies
fresh air to a naturally ventilated system that controls temperature
by distributing cool and warm air through the structure and the
radiating ceiling to a large central three-dimensional open space at
the heart of the building (Figure 12).

13 CRAB with AKT II, drawing studio, Arts University Bournemouth, UK, 2015. A smooth, single-surface structural form with five openings, producing different lighting conditions within the workspace.

13

14 Pernilla & Asif with AKT II, Coca-Cola
Beatbox Pavilion, Olympic Park, London, 2012.
The structure is more than a pavilion or
an installation, it is an interactive 'musical
instrument'. It is based on a reciprocal
geometric arrangement of interlocking ETFE
pillows into a seemingly organic, but regular
three-dimensional tiling pattern.

14

Building elements are no longer treated as separate because it is now possible to model (and therefore conceive of) buildings holistically within three-dimensional interoperable models that incorporate form with material properties and structural geometry with environmental parameters. The convergence of multiple functions within integrated building components has generated new efficiencies. For example, the close relationship between structure and envelope in an integrated whole has allowed us to conceive of structural skins and deep environmental and responsive envelopes as opposed to traditional facades.

In the Arts University Bournemouth drawing studio with Peter Cook's CRAB (Cook Robotham Architectural Bureau), the structure is the envelope of the building, fabricated by a ship builder as a doubly curved structural monocoque shell; using the latest automated processes, it is assembled in large sections, then bolted and welded together into a seamless monolithic object (Figure 13). Delamination of the facade from the inner building structure has traditionally been the clear expression of the decoupling of disciplines. The very word 'facade' signifies front, and is associated with the appearance of a building, its image. In isolation, a facade can only respond to climate through the type and quality of its materials; but today, the performance of the envelope and its integration with the inner core of the building is critical to the efficiency and sustainability of the building throughout its life. Envelope and structure, their form and materiality, work in union to produce the environmental performance of a building as a 'living organism'. From 'breathing' adaptive envelopes to radiating slabs, to the very internal and external form of buildings, all contribute to an efficient building and its appearance.

A research-based approach to practice requires an investment of resources to deliver value, and needs to prove viable for a business. What we have found is that the return is exponential to the investment because the intellectual gain, and the knowledge and tools that are developed, are resources that are constantly reinvested and refined. This investment in research spans from projects to academia. Strategically, we have put time into small to medium projects that serve as fast-track prototypical test beds for new materials, fabrication methods, geometric and organisational systems, structural types and tools. The Coca-Cola Beatbox Pavilion, designed with Asif Khan and Pernilla Ohrstedt, for the London 2012 Olympics, is an example where a simple structural organising principle has generated a multi-experiential object of interaction that uses light, sound and touch (Figure 14). The speed and simplicity of these projects allows a more immediate and instant feedback, accelerating the use of certain technologies on larger projects. We will demonstrate this through the book, showing how some of these micro projects have had an impact on decisions at the macro scale on specific designs, sometimes allowing the emergence of new market-changing typologies. The aim of our active involvement in academia is twofold: on the one hand, it engages a new generation of designers in some of the challenges that they will tackle in practice, armed with some of the knowledge

15 AKT II genealogy of structural envelopes.
A map of the evolution of structural envelopes
from the end of the last century to the present
day, classified into grids (constructed by
the assembly of discrete linear elements),
continuous rigid shells (in stress skin,
monocoque construction or cast in situ) and
tensile membrane (in fabric, stress skin or fabric).

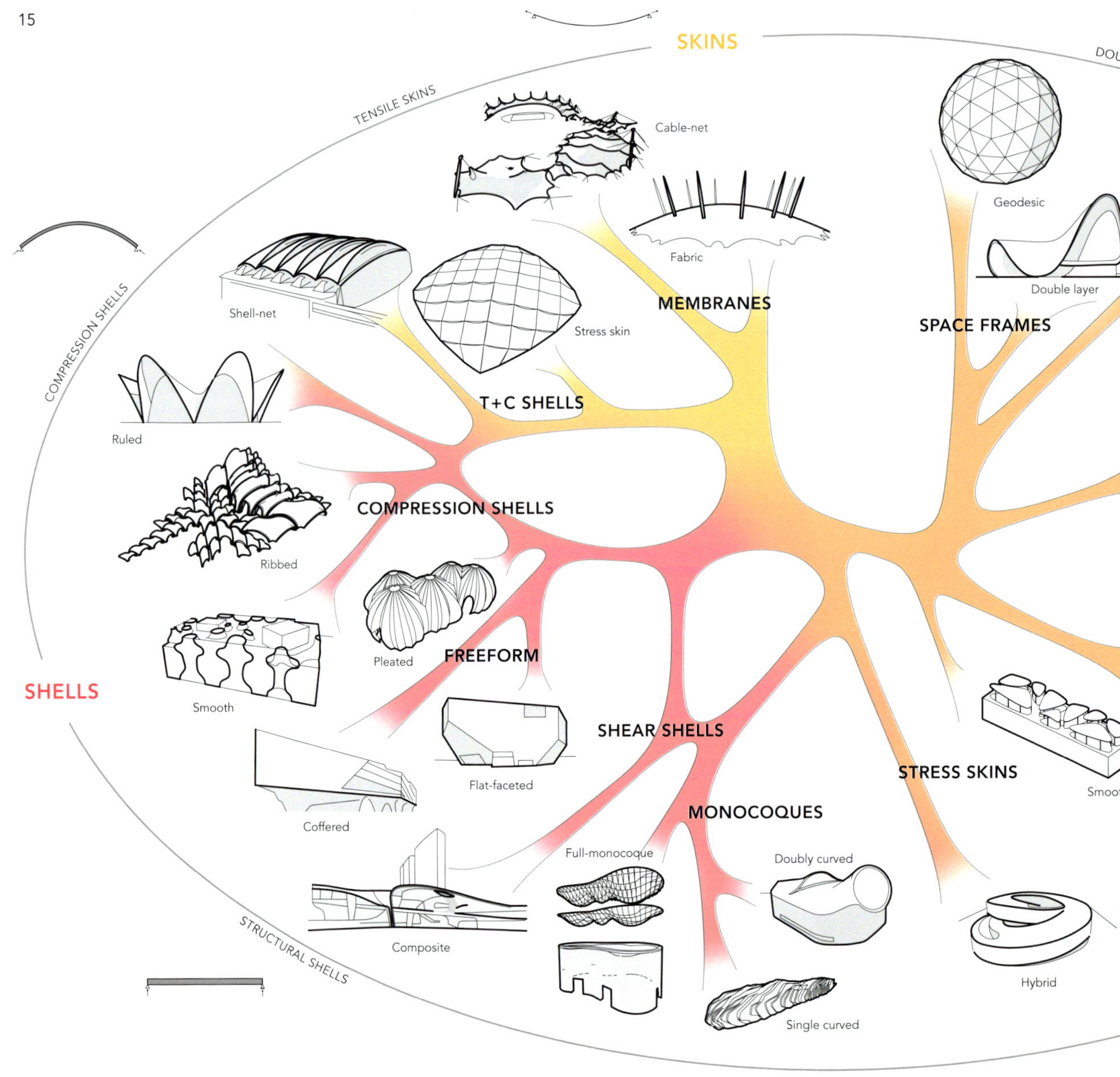

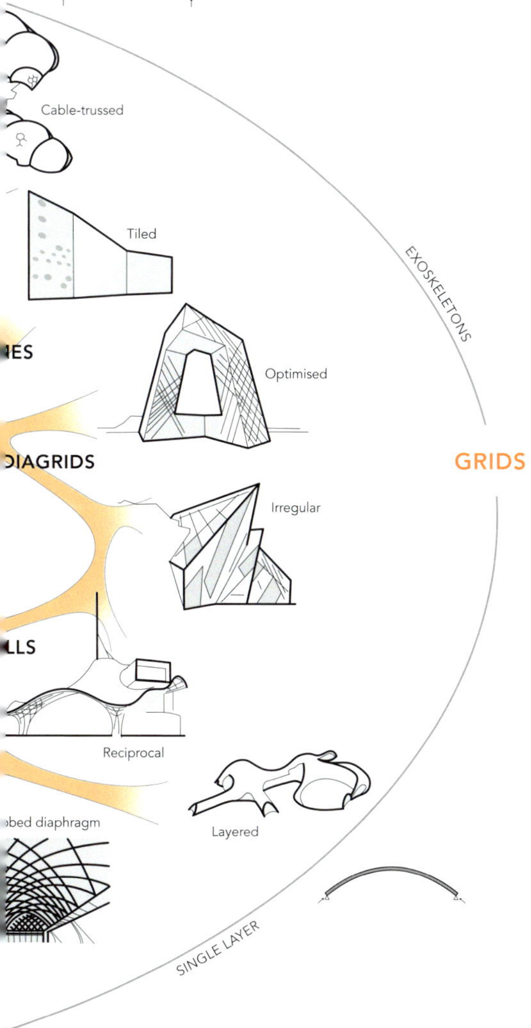

we have developed as practitioners; on the other, it exposes our work in practice to new ideas and establishes a dialogue between our generation of designers and those of the future.

Design engineering today has become a more mature profession; it has developed the tools and processes to research and apply technology in a more agile, integrated and meaningful way, while also developing the language and conceptual framework required to engage actively and deeply in a dialogue with architecture and other disciplines. With a nimbler and more dynamic way of communicating its core values of simplicity, beauty, efficiency and economy, it has been able to engrain these in the very concept and aesthetic of a building. Design engineering in the past two decades has refocused its attention from the mere art of problem solving and the mechanistic application of analytical processes to the enhancement of human well-being, and the sustainability of the environment through ingenuity and invention. It is through work with various organisations that we have been able to push the boundaries of technology, while helping them to develop new models of collaborative working through an architecture that is as much experiential as it is performative, and as focused on the individual as it is on its surrounding environment and community.

REFERENCES
1 In mathematics, a Seifert surface (named after German mathematician Herbert Seifert) is a surface whose boundary is a given knot or link. Such surfaces can be used to study the properties of the associated knot or link. These are non-orientable surfaces whose top surface continuously flows into its bottom surface.
2 A non-uniform rational basis spline (NURBS) surface is a mathematical surface used in computer graphics for generating and representing curves and surfaces. It offers great flexibility and precision for handling both analytic and modelled shapes.

TEXT
© 2016 John Wiley & Sons Ltd

IMAGES
Figures 1 and 2 © Arup; figures 3, 5, 7 and 8 © Arup/Daniel Bosia; figure 4 © Ulrich Rossmann/Arup; figures 6 and 14 © Hufton + Crow; figure 9 © Courtesy of Zaha Hadid Architects; figures 10, 11, 12 and 15 © AKT II; figure 13 © Valerie Bennet/AKT II

3 SO DIGITAL, IT'S ANALOGUE JORDAN BRANDT

Like the vanishing pixels of exponentially increasing resolution, discretisation in both computation and physical output is approaching the infinitesimal to such a degree and so rapidly that, for human observers, it almost feels analogue again. Our computer-controlled machinery and new additive technologies are on the threshold of manufacturing curves and complex surfaces, without a hint of the digital ornament that was so ardently post-rationalised only a few years ago. In some industries, such perfection has spawned nostalgic movements, driving audiophiles back to the warmth and fuzziness of needles scratching vinyl – atoms bouncing on atoms. In contrast, the second decade of avant-garde architecture in the 21st century continues its loving embrace of digital tools – a deserved, if belated, intimacy. This strengthening relationship has yielded heretofore unimaginable buildings, yet success varies by artist. The awkward shuffle of mouse clicks and keystrokes operating in a black box often foil the architect who aspires to fluid, continuous choreography. Rarely does production perform as well on the built stage as it does on the screen.

There are many unsung heroes of the computational revolution in architecture, many of whom operate away from brand-name firms and schools that make headlines. Although a minority, this cadre of fabricators and engineers from many different industries helped architects to pioneer the first wave of digital authorship in design and construction. These heroes will be called upon again to guide design intuition as a post-digital probability wave begins, pushing us to the binary threshold, wherein the physics of computing may soon arise from the same analogue phenomenon it is intended to simulate.

CONTINUITY DECONSTRUCTED

It is rare that an architect walks into a structural engineer's office and says, 'I have a differentiable equation that defines my design', but that is precisely the logic that accompanied George Legendre and his firm IJP in 2005. In a project to cover a portion of Regent Street and a neighbouring alleyway in central London, the Glasshouse Street Canopy is a transparent cover that protects pedestrians from the elements, and provides a visual symbol of the redevelopment taking place around Piccadilly. The streetscape

$$\text{Bend}_{in,jn} := \sin\left(\text{beginning\_of\_interval} + \frac{in}{\text{last\_in}} \cdot \pi \cdot \frac{\text{fraction\_of\_trigo\_arc}}{100}\right) \cdot \text{inflection\_of\_segment}$$

$$\text{slopeA}_{in,jn} := in \cdot \frac{\text{deltaH} \cdot \text{fraction\_of\_trigio\_arc}}{100 \cdot \text{last\_in}}$$

$$\text{jThread}_{in,jn} := \text{mirror\_surface} \cdot \text{mirror\_pockets}\left[\sin\left[\frac{jn}{\text{last\_jn}} \cdot \left(\pi\,\text{depth\_of\_pocket}\right)\right] \cdot \text{width\_of\_segment} + \text{mirror\_pockets} \cdot \text{Bend}_{in,jn}\right]$$

$$\text{iThread}_{in,jn} := 1 \cdot \left(\frac{in}{\text{last\_in}} \cdot \text{length\_of\_segment}\right)$$

$$\text{shape}_{in,jn} := \left[\sin\left(\frac{in}{\text{last\_in}} \cdot \text{number\_of\_pockets} \cdot \pi\right) \cdot \left(\cos\left(\frac{jn}{\text{last\_jn}} \cdot \pi\right) + -1\right)\right] \cdot \frac{\text{height\_of\_segment}}{2} + \text{slopeA}_{in,jn}$$

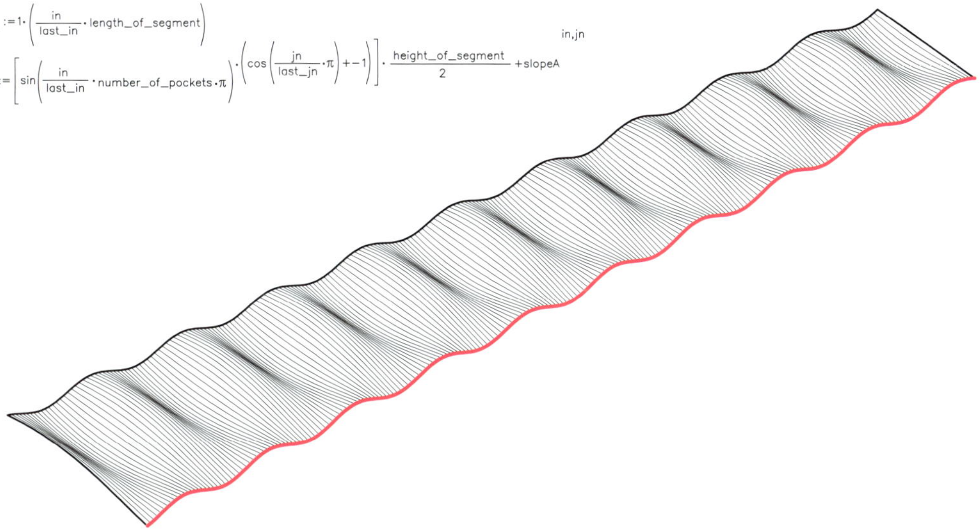

1 George L Legendre, mathematical formula and geometry for Glasshouse Street Canopy design.

of historical buildings and planned modern structures led to the canopy's integration with the new masterplan, producing a 'fulcrum' around which new projects would flourish.

The hybrid design team of architects and engineers explored several families of analytically defined surfaces, reminiscent of flowing waves and aperiodic sinusoidal functions, before deciding to focus on an 'asymmetrical weave' concept. By manipulating a few parameters in a set of explicit mathematical equations, the team was able to articulate an intriguing form that seemed plausible for construction. The modules were 4,800 mm wide, and the interstitial 'fishbone' skeleton could accommodate larger structural members that would carry loads to vertical supports and the perimeter buildings. This straightforward logic, buoyed by a multitude of curvature analyses and preliminary hand calculations, suggested that the complex doubly curved form shouldn't be too difficult to build.

One would think that if the beginning of a design is a clean formula, then it should be easy to provide frictionless input for sophisticated structural analysis software, but these assumptions were a lost cause. After routing the geometry through several software packages, it was ultimately decided to write custom scripts and port the formula into a format legible by the analysis package, resulting in an approximated, discrete interpretation of the structure.

Constructability concepts were developed in parallel, adding to the growing repertoire of subdivision algorithms, and generating

2

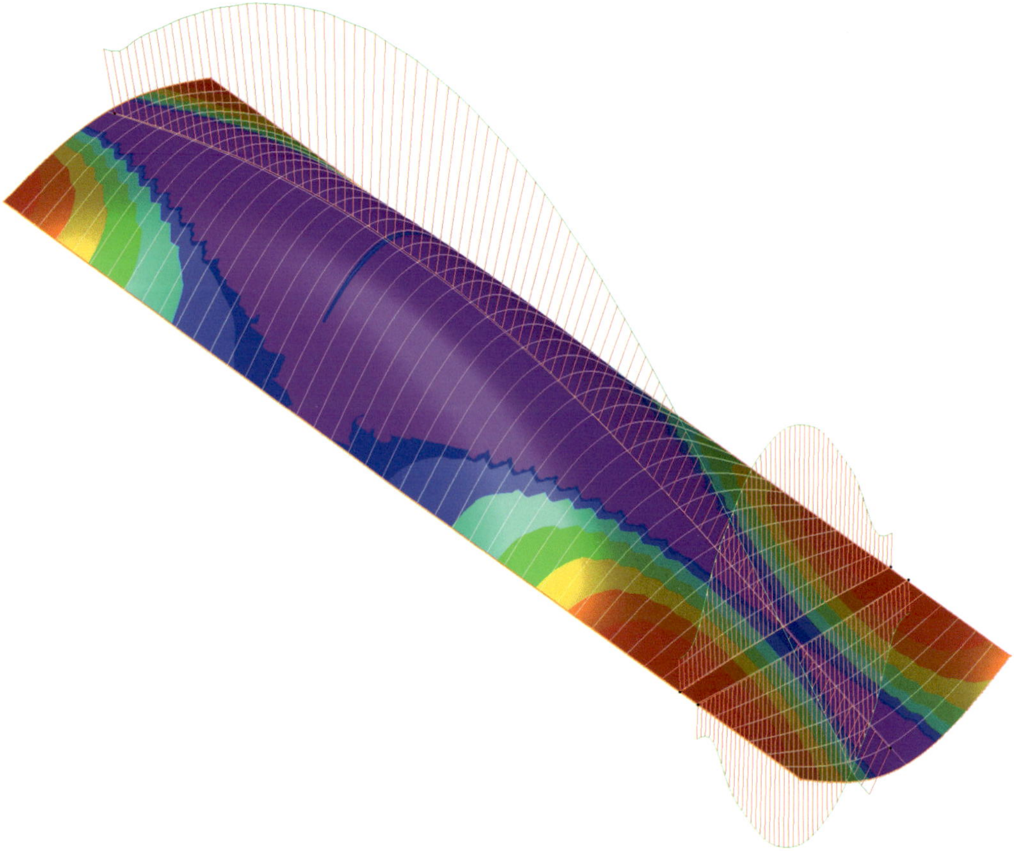

3

several sets of detailing options that would allow for cheaper, planar panes of glass that could be used to approximate the form within a reasonable deviation of the mathematical surface. Even by the mid-2000s, this had become a recurring theme with at least two decades of precedents from which to learn. The Great Court renovation at the British Museum remains a strong example of a contribution to an evolving computational craftwork to solve modern incarnations of formal complexity. Nevertheless, as is the case in architecture, each instance exposes specific nuances that must be addressed, so custom details inevitably emerge in every project that involves tessellating curved glass surfaces. Hence the beauty of Legendre's continuous mathematical function differentiated into the limits of steel tubes, glass panes, aluminium mullions, rubber gaskets and silicone caulking; after all, discrete components need analogue bindings.

MATURITY OF MIND AND TOOL

The sophistication of formal representation in the Glasshouse Street Canopy project by no means represents the standard of descriptive maturity at the time. Most architectural practices were still in the process of mastering digital craft, particularly in the construction of complex surfaces. As young, digitally savvy protégés presented concept models to the lead designers and followed instructions to tweak and modify the shape, successive iterations often led to patchwork geometries. Erratic inflection points and unintended singularities appeared; any mistaken mouse clicks are reflected in the geometry of the design.

Much of p.art's role in these pioneering times involved manipulating input geometries and educating designers on means and methods to generate rational form conducive to analysis and construction. At the time, it was not universally understood how to maintain continuous curves of inflection when creating forms that transition through positive and negative Gaussian curvature, or how to apply constraints on adjacent surfaces to ensure that tangent and curvature discontinuities did not manifest unwanted artefacts on the built exterior.

In return, architects educated aspiring design engineers about the reasons for which iteration is a fundamental property of design. As much as a structural mind would like quickly to freeze form into a static problem, design experience suggests that decisions should be suspended until all variables are quantified or qualified within an acceptable range. Digital computation mitigates a portion of this professional rift by powering faster cycles of design, analysis and what-if scenarios; however, the order of approximation required to conduct these studies is still a limiting factor.

Even in the early days of design computation, it was understood that increasing processing power would radically transform the industry. Now, with nearly infinite scaling through the cloud, driven by the collaborative capacity among distributed human agents from whose behaviour machine learning will fully leverage computing resources, it is hard to predict where the exponential

2 IJP Corporation with AKT II, planar glass study for Glasshouse Street Canopy design.

3 IJP Corporation with AKT II, Gaussian curvature analysis of different concept models. The model on the left demonstrates good control with (mostly) continuous inflection. The model on the right demonstrates poor control with erratic, singular inflections.

curve will lead. One thing is for certain, we will never find answers
to questions we do not ask. Beyond informing multi-objective
design challenges and accelerating iteration, what other important
problems are being solved? How many virtual nails have been
invented by this nearly infinite digital hammer?

THE MISLED SURFACE FETISH

The definition of two-dimensional objects, whether embedded in
a Cartesian framework or intrinsically parameterised exclusive of
space, has become an obsession of architectural representation.
Recent years suggest a growing trend towards discrete objects,
agents and swarms, but the surface remains; a beacon of
mathematical purity. Purity is deceptive in an applied field. If the
built form consists of atoms filling a volumetric space, then why
do we persist on an infinitely thin abstraction? The fallacy of this
paradigm is evident when investigating the interstice between
exterior surface and interior structure in any contemporary
building with some degree of complex curvature. Tertiary systems
of varying sophistication inevitably inhabit this wasted space

4

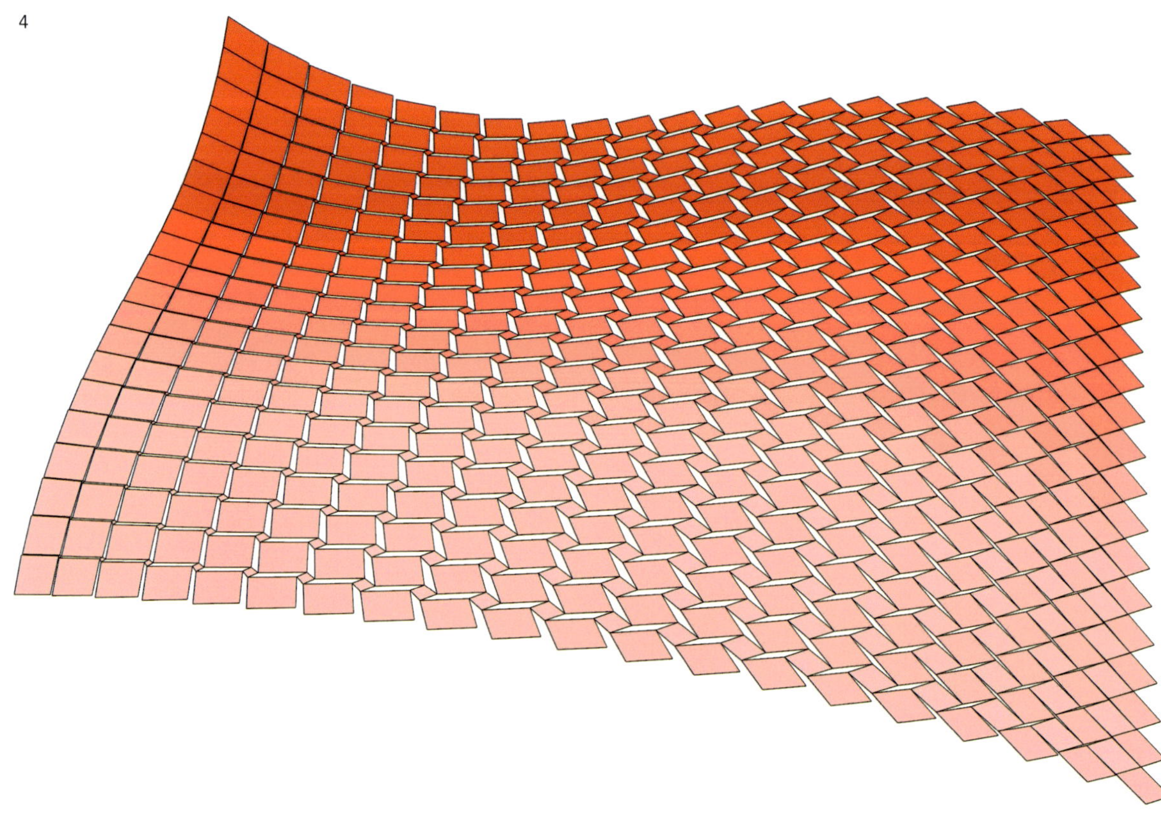

because designers too often rely on traditional contractors to define what happens between the layers: stucco sprayed on wire mesh, sticks bolted to expensive connectors on aluminium tubes, or composites built on disposable formwork. These all embody false constructs to offer an illusion of exterior continuity, the material hangover of our surface obsession. A pedagogy of formal representation, exclusively through surfaces or curves (whether as models or drawings), ensures a slow death for its disciples, particularly as new additive manufacturing processes emerge that allow us to place matter precisely where we want it. Our geometric currency and intuition of material and process must evolve, with design engineers acting as the catalyst.

If two decades of subdivided surfaces and planar tessellation studies in academia and practice are any clue, then it seems we have avoided the true constraint of constructing curvature, and settled for a digital compromise; the methods of manufacture, not the representation, have been the true challenge. Fortunately, other industries, aided by a select few in architecture, have invested heavily in new production methods so that compiling matter at the micron scale may soon be a reality. This could obviate the need to segment curves or rationalise surfaces with tertiary exterior structure altogether. However, much of the design community continues to search for nails to pound with their digital hammer.

Some problems, and their solutions, are conducive to computing in binary, which is in part why the industry now suffers from a sort of digital determinism. This stems from a belief that breadth of scope and precision of definition necessarily yields design intent in built form, from the computer to the job-site. But building information models boasting 10-digit floating-point accuracy and CNC machines operating in sub-millimetre increments do not instruct a worker on the end of a 20-metre lift squirting silicone out of a tube while it's raining. Construction is still a physical process governed by tolerances and probabilities. It may prove valuable to regain the fuzzy logic of analogue definition, that which requires designers to visualise and comprehend physical behaviour beyond their screen.

GEOMETRIC CURRENCY OF COMPILING MATTER
Solid models are increasingly common in architecture, highlighting a positive trend for the industry. However, even in sophisticated CAD/CAM packages these are merely closed surfaces that represent bounded volumes, hence the term 'boundary representation' (BREP). For traditional engineering materials that are cut, cast, formed, machined and assembled, such models are adequate, but they fall short when applied to organic, algorithmically generated designs with material gradients and trillions of machine instructions. The assumption has always been that a single material (or multiple) with a regular pattern will correlate to a digitally represented volume – an aluminium engine block or masonry wall with stone and grout, for example. Some researchers are challenging this assumption, including members of the p.art team, who have developed

4 AKT II, planar tessellation of double-curved surface.

5

6

custom software for topology optimisation and functional material distribution. Regardless of the geometric currency that prevails (voxels, tensor fields, level-sets or other hybrid formats), future designers will operate across an intriguing threshold between discrete and continuous means of representation.

In a future that includes the compiling of matter at the micron scale, there would be no need to assemble discrete components from glass, aluminium and rubber to make a watertight curtain wall. Simply compile a gradient from rigid, opaque load-bearing material to a translucent elastomer transitioning through transparent amorphous polymer, producing a single solid of many materials. Now consider how to instruct a machine that precisely places each bit of material; the design model must not only contain dimensional information, but also specifications such as colour, composition, density, etc. If each cubic micron required about 25 kilobytes of data for the definition of material properties, cellular geometry, coordinates and other metadata, then something the size and complexity of a mobile phone would require nearly 2 petabytes of data. Even with compression and a host of optimisation algorithms to reduce the size, there would still be a massive amount of generated data and additive machine code that current mental and digital models are ill-prepared to orchestrate.[1] We've become expert at differentiating the continuous, but how do we create continuity from disparate bits?

Articulating details at the micro-architectural scale would prove a formidable intellectual and computational challenge, compounded by the complexity of organic morphology observed in generative design. High genus topologies, that is to say shapes with lots of holes, become memory intensive and sluggish as a function of the operators used in boundary representation, whereas alternative voxel models and level-set methods are indifferent to such complexity. This is an important distinction, as these shapes are not a trivial manifestation of style or expression of digital prowess such as Blobs in the late 1990s. Form is derived from a computationally optimal distribution of matter and energy that solves the human problems we explore; as such it often assumes an eerily organic structure. After all, nature is just trying to find its lowest energy state. Design engineers play a pivotal role in this environment, because they challenge the architect to focus on asking the right questions instead of assuming an answer. This is like suppressing a gag reflex for most designers; when considering a problem, it's nearly impossible for them to prevent form from materialising in their mind's eye. When we read a good novel, we suspend our disbelief and commit to the author's narrative. With the power of generative design and nearly infinite computation, we must suspend our *belief* and not make assumptions about how things should look.

PARTICLES AND WAVES

In some ways our frustrated social experience of uncertainty (plans that aren't executed, events that cannot be foreseen, and people who are unreliable) lures us into the trap of digital determinism,

5 Autodesk, 3D printed multi-material cube with functional gradient from elastomer (top) to rigid polymer (bottom).

6 Autodesk, generatively designed pedestrian bridge.

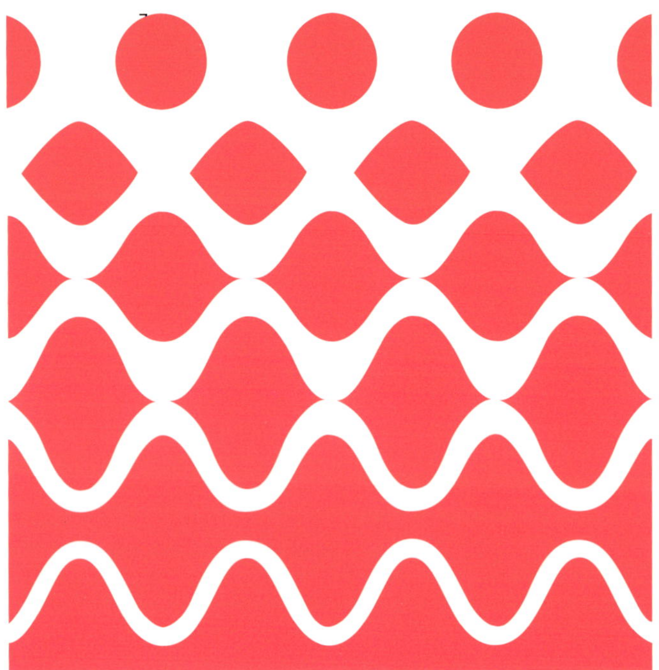

7 AKT II, artist's sketch of particle-wave duality
phenomenon.

wherein we believe that there is a single answer to our questions.
The promise of precision is seductive. Modern computers and
communication networks allow us to read, write and transmit
discrete 1s and 0s without error, masking probabilities that govern
the quantum world of physical information processing as particles
bump into each other. When observing the massive revolutions
that this fairly recent digital phenomenon has fuelled across all
aspects of human existence, architects and engineers can't be
blamed for being similarly seduced. It has, without question,
changed the nature of both the art and practice, but even bigger
changes lurk in the continuous threshold of grey area between 0
and 1. We have become so digital, that we're going analogue.
Quantum computers, essentially analogue machines that specialise
in manipulating quantum particles and asking them in which
state they are likely to be found, offer a path to simulate physical
reality using physical reality. Instead of floating-point accuracy,
think of fuzzy-point probabilities. Formulate analogue questions
as qubits with a superpositional state between yes and no rather
than polarising continuity into this or that as we do in classical
circuits. Qubit logic aptly translates to the optimal distribution
of materials and myriad other mechanical problems. Rather than
declaring that: 'a discrete area of space should be occupied by
this or that material', we can ask: 'what gradient of materials can
best solve the set of multivariable functions we define in a spatial
continuum?' Questions remain in superposition until the wave
function collapses into a definite state, a probabilistic answer, when
it is measured.

How will industry adapt? How will bankers and insurance
underwriters react to an architect stating that: 'it is 84.321933%

likely that this building will meet the customer, regulatory, environmental and financial requirements'? Want something better? Roll the dice a billion more times per second and see if chances improve, or go back and rephrase the questions. It's a good policy against premature optimisation.

The collective future of design operates at transition from digital to analogue, foretold by our early-20th-century scientific forebears who had an uncanny notion that things fundamentally behave in unintuitive ways. Cultivating a post-digital intuition becomes an opportunity for design engineers. The qualities they display continue to expand: proficiency in an increasingly broad technical repertoire, qualitative sensitivity and the capacity to articulate creative inspection. The questions that we ask and how we ask them become central to the applied architectural enterprise, while obsession with representation and craft will commoditise myopic firms. Thus the values of p.art suggest leadership over analysis, experimentation not assumption, and the art of inquiry between yes and no.

REFERENCES

1 One possible outcome is that designs are compiled simultaneously with the product. The machine code would be created by generative algorithms as the information is necessary to make the 3D print or to run analysis. A complete design 'file' may never exist for unique products built from custom computational materials, only the digital DNA.

TEXT

© 2016 John Wiley & Sons Ltd

IMAGES

Figure 1 © George L Legendre/IJP; figures 2, 3, 4 and 7 © AKT II; figures 5 and 6 © Autodesk, Inc

4 ARCHITECTURE–ENGINEERING INTERFACE

SAWAKO KAIJIMA AND PANAGIOTIS MICHALATOS

With the proliferation of new technology, it is possible to bring existing industry structures and cultural practices into question, providing opportunities for their redesign, particularly that of the architecture–structural engineering relationship.

The separation of the two disciplines can be traced to around the time of the Industrial Revolution, marked by its multiple technological developments, and the gap has been growing ever since. This increase has been institutionalised and exacerbated by the divergent tools, methods and conceptual frameworks that practitioners of engineering and architecture are exposed to.

Now, with the intervention of digital technologies in this relationship, the underlying data structures and representation of both fields are transforming. A side effect of digitisation is that different objects become commensurate, allowing disparate methods to be transferred and applied to multiple fields. In addition, the degree of communication and synergy between practitioners in both architecture and engineering has intensified: therein lies an opportunity for computational methods to renegotiate boundaries, fill gaps and facilitate overlaps between the two fields, in effect designing a new interface.

Another effect of digital design tools is that design software creates the belief that form can be conceived in a vacuum, domain product of a solipsistic design intuition: this is the 'free-form' paradigm. Architects can design 'whatever' and engineers will do 'whatever' it takes to build the object in question, an approach that Eduard Sekler would have characterised as atectonic.[1] While the incredible technological advancements in post-rationalisation and digital fabrication which make such an approach viable can't be dismissed, this solo approach forgets the biggest promise of the information revolution: increased communication, connectivity and interoperability. This latter aspect is what the idea of a design interface seeks to address, and it highlights the fact that architectural projects are developed by multiple agents.

Within such a system, the relationship between structure and form is not simply a question of efficiency or its opposite, and it is complicated by two factors. First, the optimal solution for real-world problems is never a single one, simply because the environmental context is transient and unpredictable, boundary

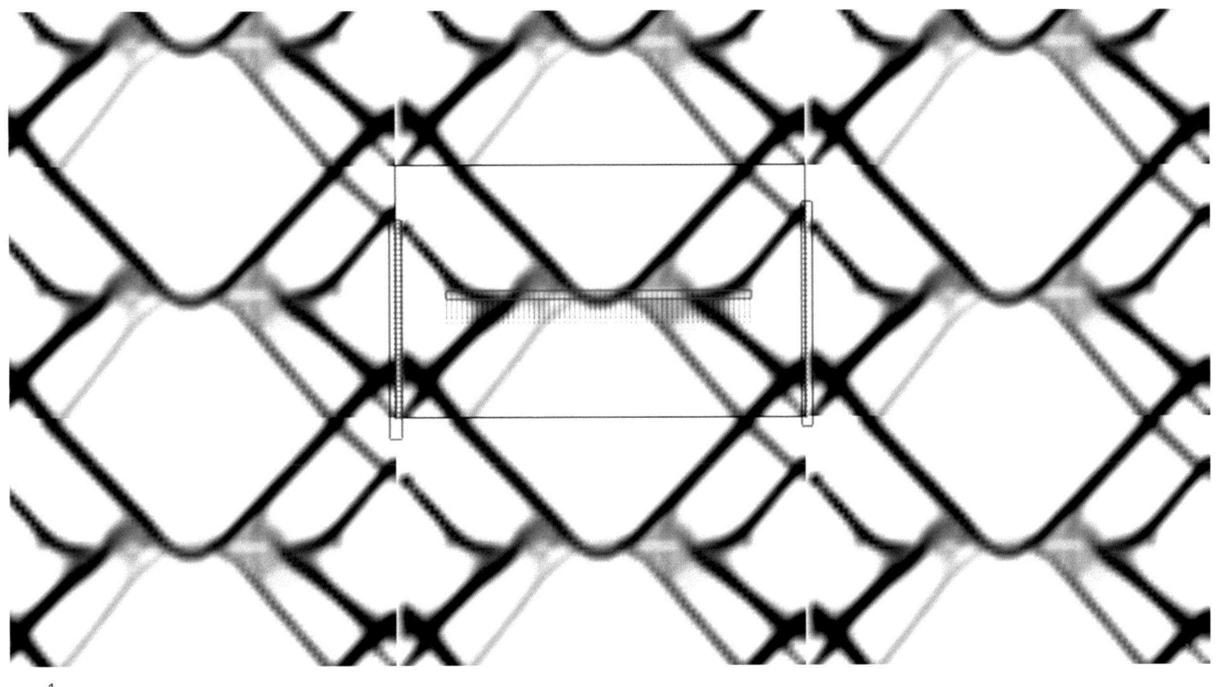

1

1 Topostruct, 2008.
Screen capture of custom software, Topostruct,
showing a result of Periodic Topology
Optimisation.

conditions unknowable, and goals hard to define. Second, even the most utilitarian object produced by the most utilitarian mind will become an object of aesthetic judgement, interact with cultural forces, and be invested with meaning as soon as it is placed in public view.

Digital media, using a language of efficiency and optimisation, has been used to address the relationship between structure and form to confront the first complication only, but throughout architectural history this has always had implications for expression and meaning. The degree and mode in which structure is expressed or challenged in the architectural form can fluctuate from indifference and structural obesity of hidden structure, through structural exhibitionism and hyper-efficiency, to the other extreme of radical thinness, structural atrophy, and a desire for the disappearance of structure altogether.

In light of computational technologies, the Computational Design Consultancy was formed at AKT II in 2006, serving to interface architecture and structural engineering. In his book, *Interface*, Branden Hookway,[2] describes interface as a process of drawing incompatible entities together to create compatibility, which involves the consideration of technical, social and perceptual aspects, among others: '… Interface is both the bottleneck through which all human relations to and through technology must pass, and a productive moment of encounter embedded and obscured within the use of technology.'

The work of the Computational Design Consultancy team from 2006 to 2010 included three categories, within which considerations around interface were subconsciously addressed and worked through. These three categories are:

- · Traditional interface: taking advantage of the common language.
- · Communication interface: navigating through the issue of control.
- · Intuitive interface: computer–human interface for design.

TRANSLATION INTERFACE: TAKING ADVANTAGE OF COMMON REPRESENTATIONS

Today, most parties related to the production of architecture utilise various software tools to perform many of their tasks. For instance, architects use 3D modelling tools to simulate architectural geometries, while engineers use finite element (FE) software[3] to simulate structural behaviours. Though software tools generally try to facilitate production within each discipline, they do share a common underlying representation of information in a digital format that can be accessed by using programming languages. This is significant for the technical/technological communication channel between architecture and engineering.

Automation: DRL 10 Pavilion

DRL 10 is a small pavilion designed to mark the 10th anniversary of the Design Research Laboratory programme at the Architectural Association. The geometry for this project was designed in Rhinoceros as a doubly curved surface, intersected with a series of planes to generate the structural ribs.

The role of the Computational Design Consultancy was as an internal service that assisted engineers in building FE analysis models in SOFiSTiK[4] from the architects' models in Rhinoceros.

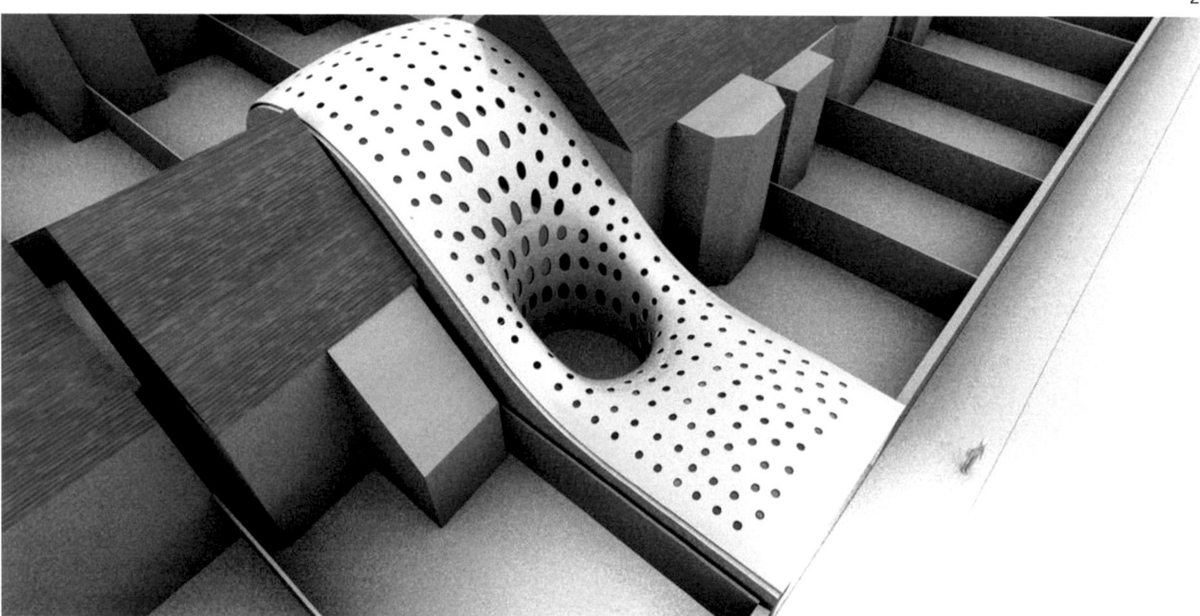

The translation of models between software tools is one of the fundamental obstacles in developing integrated solutions to a project that yields problematic conditions at various levels; one such problem is the time it takes for the conversion process. This is particularly evident when faced with projects involving complex geometry, since the translation of models tends to take a lot of time away from both architects and engineers. It is also worth noting that model translation is not a direct process; extra information has to be added during the process.

The DRL 10 Pavilion demonstrates the effect of minimising model translation time because of its challenging geometry, material use and short time frame. By developing a project-specific conversion software tool, it was possible to respond quickly to requests from the engineers in testing the structure: more than 100 FE models were produced for the project in just a few weeks, which would have been impossible otherwise.

Though most work of the computational design team involves facilitating communication between engineering and architecture software tools, for this project the main focus was on speeding up the traditional workflow, based on a continuous trial-and-error loop and intuitive decisions. The large number of models required emphasises the slow convergence of such procedures.

Structural expression over geometric substrate: Clyde Lane House
The Clyde Lane House (Figures 2 to 4) project comprised a roof designed by Future Systems as a doubly curved surface that dips in the middle to form an atrium. Although the roof is load-bearing, its form is not structural (structural efficiency was not a driving force behind its geometric form). For this roof design, the main communication took place among the architects, engineers and the computational design team.

The initial concern for the architects was to distribute the roof openings to account for the lighting requirement derived from programmatic constraints. The first step of any project is to render such architectural concerns suitable for computational representation, as well as developing a computational solution that is generic enough to accommodate considerations and contingencies that may arise in later stages. In this case, the main problem was the density and directionality of the opening distribution pattern.

Additionally, an extra layer of information was added to the opening pattern so that it expressed the structural logic of this 'unstructural' surface. For this, the computational design team opted for an algorithm capable of generating a mapping/ reparameterisation of any surface topology that complies to given scaling and directional conditions expressed as a pair of orthogonal vector fields. A good candidate for the input vector field, from an engineering perspective, was considered to be the two principal stress eigenvectors taken from an FE analysis of the given surface. This field was chosen on the assumption that for

2 Future Systems with AKT II, Clyde Lane House, Dublin, Ireland, 2008.
Clyde Lane House roof designed as a doubly curved surface with perforations.

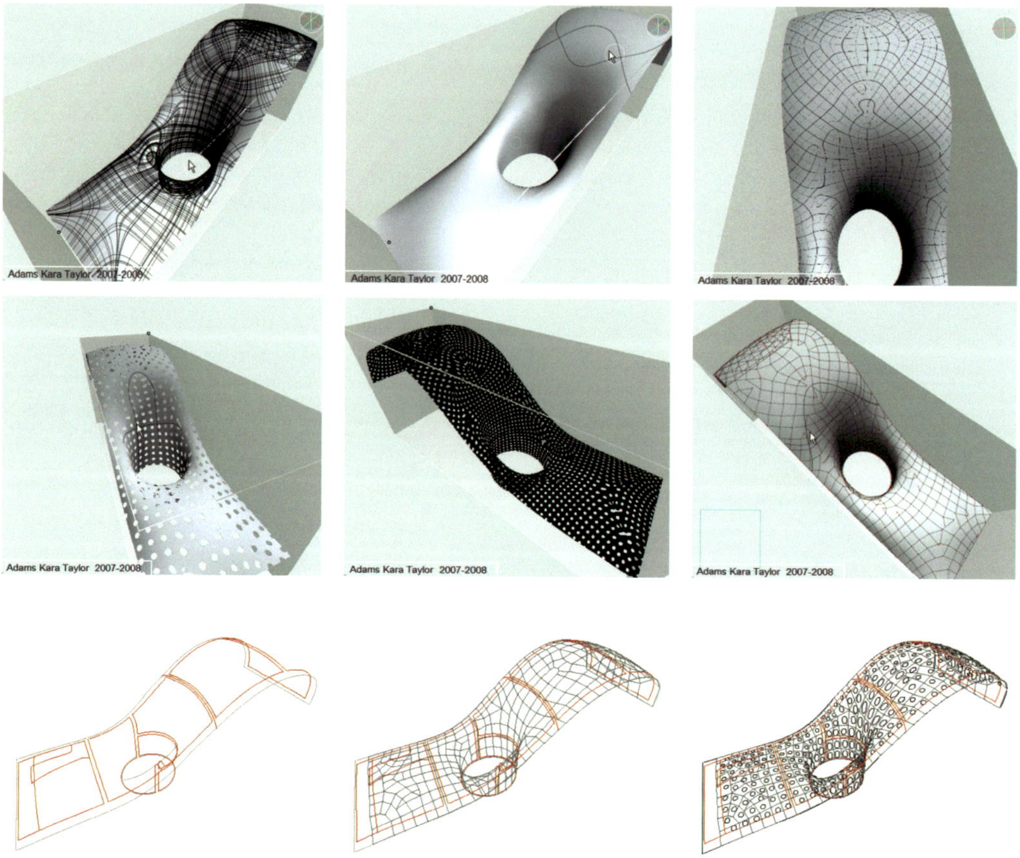

3 Structural patterns, 2008.
Research showing different
opening distributions over a
slab and its effect on principal
stress lines.

large enough surfaces with a fine material continuity, preservation of the material continuity along the principal stresses will result in some structural efficiency, while at the same time rendering the pattern as a legible sign of the underlying structural behaviour. In order to be able to locally control scaling of the pattern, an algorithm described by N Ray, WC Li, B Lévy, A Sheffer, and P Alliez, called Periodic Global Parameterisation, was employed.[5]

The project faced three major contingencies that led to formal changes: a change of opening geometry, a change of global surface geometry due to planning requirements, and the introduction of panel joints due to transportation issues. This was made possible computationally, because the most generic solution possible was selected.

COMMUNICATION INTERFACE: NAVIGATING THE ISSUE OF CONTROL
One of the difficulties of interfacing exists because of the gap between what is technically and technologically possible, and what is culturally acceptable. Projects like Clyde Lane House have utilised sophisticated algorithms to achieve integrated design between architecture and engineering considerations. These efforts were not well received, however, as it seemed that design control had been taken away from both the architects and the structural engineers. This highlights the political landscape of the building industry, and presents interface as an issue of control. In subsequent projects, communication across design and engineering team members became the focus, rather than just proposing a technical solution. In other words, the projects concentrated on making interface near-invisible, while inviting people to participate in the computational design culture.

Cross-disciplinary observables: UK Pavilion, Shanghai Expo
The UK Pavilion for the Shanghai Expo 2010 (Figures 5 to 7) was a project designed by Heatherwick Studio and structured by AKT II. The pavilion is composed of 60,000 slender 7.5 m long rods, made of transparent polycarbonate material, which perforate each side of the structural box. The unique interior is formed by the end points of these equal-length rods, smoothly distributed on a filleted offset cube.

The computational design team designed an interface for the early stages of design to observe and discuss the properties and problems of different distribution methods for the rods, seen both as individual elements and as a continuum with statistical and smoothly varying aspects. This helped to identify that a satisfactory distribution of interior rod ends would generate clashes at the corners of the structural grid of the box, and this was communicated to the team. An intuitive understanding of this potential problem, together with an understanding of the rod distribution logic, helped the architects to develop a solution that negotiated between their design objective and the positioning of the structure. At the same time, the engineering side was able to develop a subtle solution to accommodate any remaining clashes by applying a simple curvature to the structural ribs.

4 Future Systems with AKT II, Clyde Lane House, Dublin, Ireland, 2008.
Field construction and roof opening distribution design using custom software for surface reparameterisation.

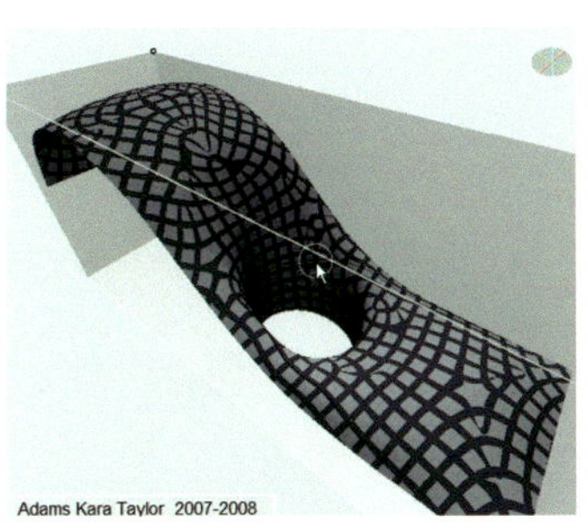

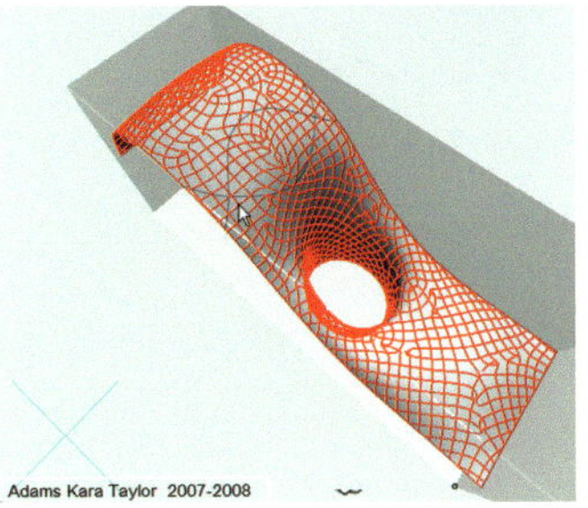

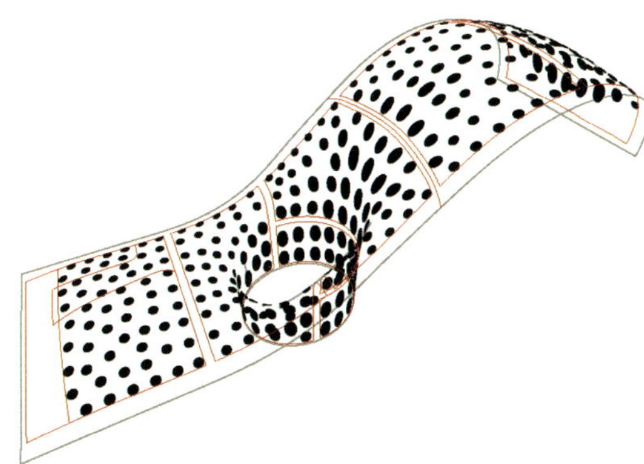

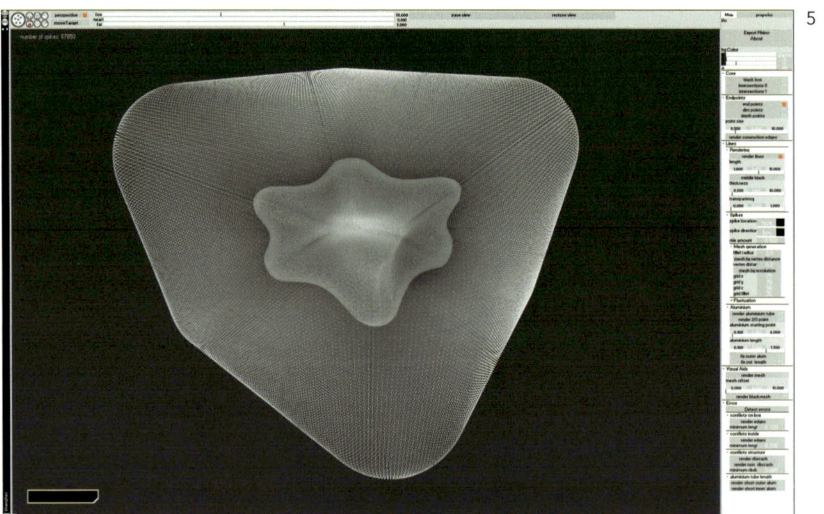

5

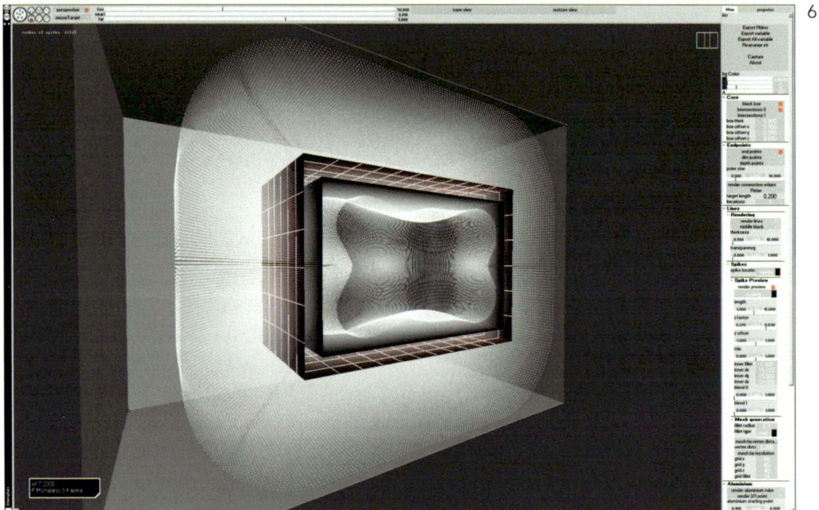

6

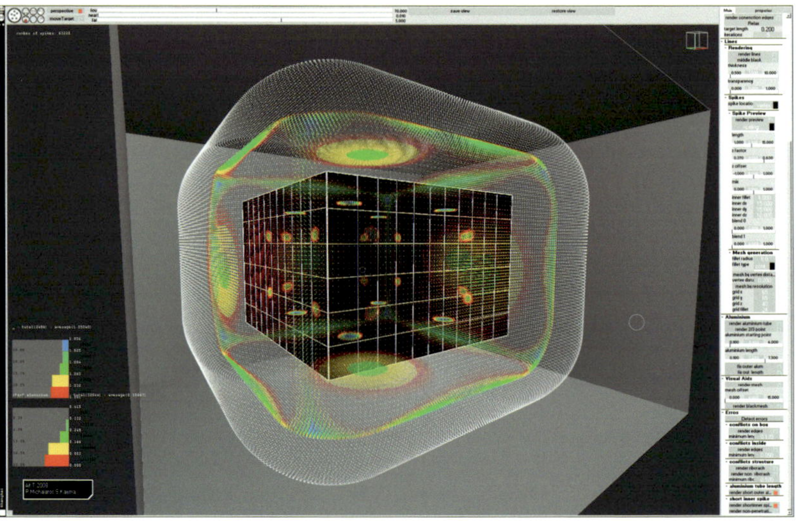

7

The tools were designed so that they would not just yield a singular optimal solution, but would allow the team to interactively develop their intuition into specific design problems in relation to questions of structural efficiency and expressiveness. This ability to operate intuitively at the crossroads of the discrete and the continuum is critical for projects that are geometrically hyper-fragmented and that work in multiple scales.

Exploring the implications of design decisions: Heydar Aliyev Center

The Heydar Aliyev Center in Baku, Azerbaijan (Figures 8 and 9), was designed by Zaha Hadid Architects. The roof is a series of ribbon-like surfaces, with a total area of more than 40,000 m^2. For this project, the Computational Design Consultancy primarily stood between the architects and the fabricators during the early stages in order to assess the external roof cladding, which is composed of a large number of distinct panels.

This required an analysis of the overall surface, as well as the properties of individual panels, in order to estimate the relation between design decisions and cost. Handling a large amount of discrete information is trivial in computation, but communicating the correlation between complex parameters to the whole project team is one of the most challenging aspects of the Computational Design Consultancy's work. This is partly because the various parties involved in production tend to have different modes of communication: visual communication is better suited to architects, whereas numerical descriptions and spreadsheets work better with engineers. Developing a software tool that reveals the visual effects of numerical input, and vice versa, helped the involved parties better understand the correlation between the design alterations and panel cost, and facilitated the decision-making process.

INTUITION INTERFACE: COMPUTER–HUMAN INTERFACE FOR DESIGN

Computer–human interface is another critical area to consider when discussing the architecture–engineering interface through computational technologies. The use of technology requires humans to participate in a particular condition, way of thinking and operation. Careful design of the computer–human interface is necessary for technology to become a means of productive augmentation for humans, and more specifically for designers.

Effect as design input: Guggenheim Vilnius

The focus for the Guggenheim Vilnius project (Figures 10 to 13) was the development of an optimal roof-opening solution that would satisfy the internal lighting conditions in a way that remained environmentally efficient throughout the year, and that would fit inside the required structural depth.

In many cases, specific information is required for developing solutions that are not possible to mediate through descriptions and text alone. The desired lighting conditions from the architects were difficult to communicate, as the architects were constantly changing the design. In order to establish a flexible input system

5 Heatherwick Studio with AKT II, UK Pavilion, Shanghai Expo, China, 2010.
Screen capture of custom software illustrating the internal organic space created by 60,000 rods of equal length.

6 Heatherwick Studio with AKT II, UK Pavilion, Shanghai Expo, China, 2010.
Sections and spike projection studies.

7 Heatherwick Studio with AKT II, UK Pavilion, Shanghai Expo, China, 2010.
Spike distribution and error visualisation using bespoke interactive software.

8 Zaha Hadid Architects with AKT II, Heydar
Aliyev Center, Baku, Azerbaijan, 2012.
A screen capture showing cladding panel
analysis and cost estimation using bespoke
interactive software developed for the project.

for the architects to communicate this information, a minimal-input interface was developed. In the past, it would have been difficult to engage people with no computation background when distributing complete software that exposes all functionalities: complicated interfaces and the underlying algorithms seem to confuse people and decrease their interest in taking such a path. For this reason, the interfaces used by other parties are minimal, user friendly, and implicitly ask for the very specific information required to generate the design solution. Therefore, two sets of interfaces were developed in parallel: one for the architects to input, and the other for the computational design team to develop the solution and the interface necessary to communicate the relationships between the input parameters and output.

This project was successful in communicating the possibilities and the flexibility of the solution to the team, since the simple and clear request for information was efficiently disseminated to the involved parties without distracting them with technicalities.

Shifting paradigm, from assembly to distribution – Topostruct
Topostruct (Figures 14 to 17) is one of the applications developed by the Computational Design Consultancy, and is based on the theory of topology optimisation.[6] This is a well-documented methodology that produces notionally optimal geometric and material distributions with respect to structural behaviour.

Topostruct was made for exploring lateral structural solutions, but also for enhancing a designer's intuition into structural behaviour. In this method, a volume of variable density or porosity material is assumed in space, and the problem is set up as a series of support conditions and forces acting on this volume. Iteratively, the related algorithms attempt to find a topologically optimal solution of material distribution that best meets the given boundary conditions, while minimising material use. To enhance the understanding of the structural behaviour of the solution, it offers different ways of visualising the information in 2D and 3D.

The software has allowed people with little prior knowledge of engineering to acquire an induced understanding of the mechanics of material. In fact, after spending time 'playing' with hypothetical scenarios and observing Topostruct, users tended to anticipate the results, as though they had gained an intuitive understanding of the underlying principles.

Interfaces as paradigms
In effect, designing an interface is a way to redefine the form of communication and, at the same time, affect the vocabulary of the discourse and the internal thought process of the designer. It is erroneous to see software applications as just another tool. Software applications present a curated view of the design problem and determine what is observable and significant, as well as the allowable actions and possible criteria of evaluation. In a sense, they constitute environments, rather than just tools. Such applications partition the continuum of formal manipulation into a

9 Zaha Hadid Architects with AKT II, Heydar Aliyev Center, Baku, Azerbaijan, 2012. Fabrication information and panel rationalisation over a series of ribbon-like surfaces, with a total area of more than 40,000 m².

9

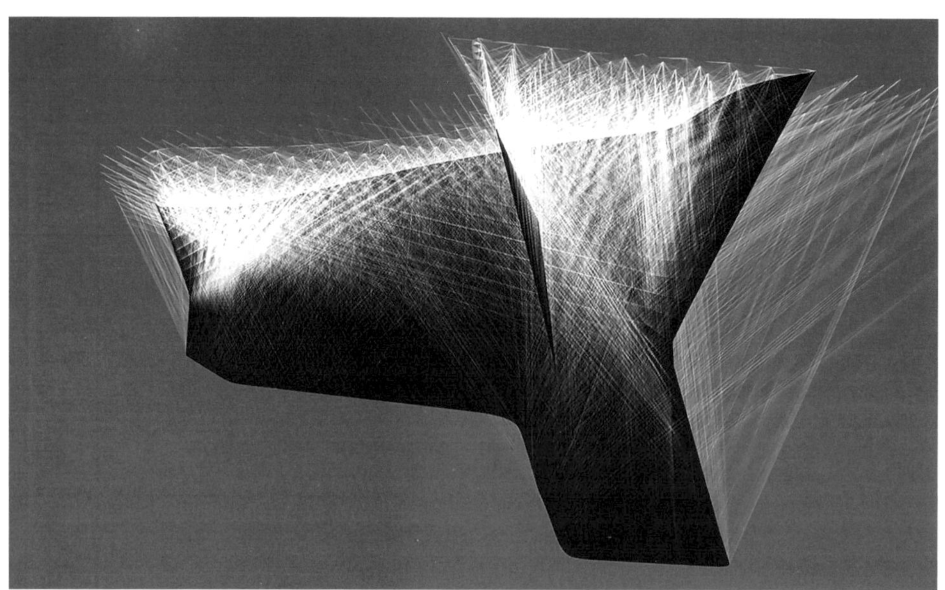

10

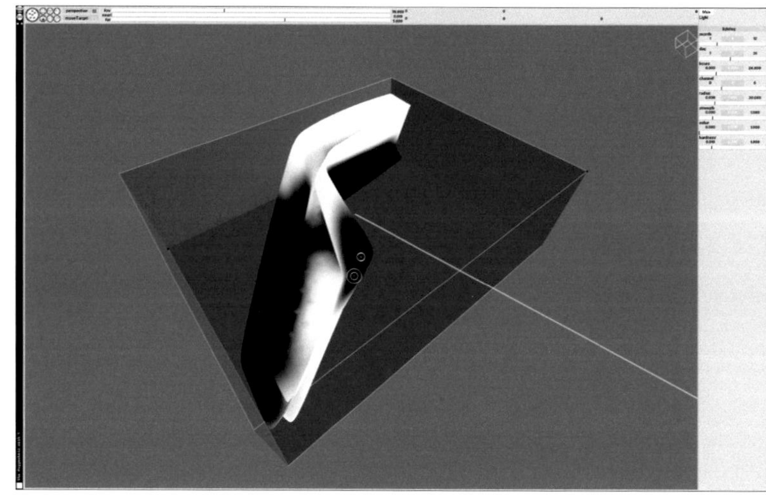

11

12

10 Zaha Hadid Architects with AKT II, Guggenheim Hermitage Museum, Vilnius, Lithuania, 2007.
Visualisation of the inverse illumination process used for designing the roof openings of the Vilnius museum and cultural centre.

11 Zaha Hadid Architects with AKT II, Guggenheim Hermitage Museum, Vilnius, Lithuania, 2007.
A simple painting interface to describe desired lighting conditions. The painted mesh geometry was used as design/data input for the subsequent inverse illumination process.

12 Zaha Hadid Architects with AKT II, Guggenheim Hermitage Museum, Vilnius, Lithuania, 2007.
Illumination analysis of the optimised roof openings.

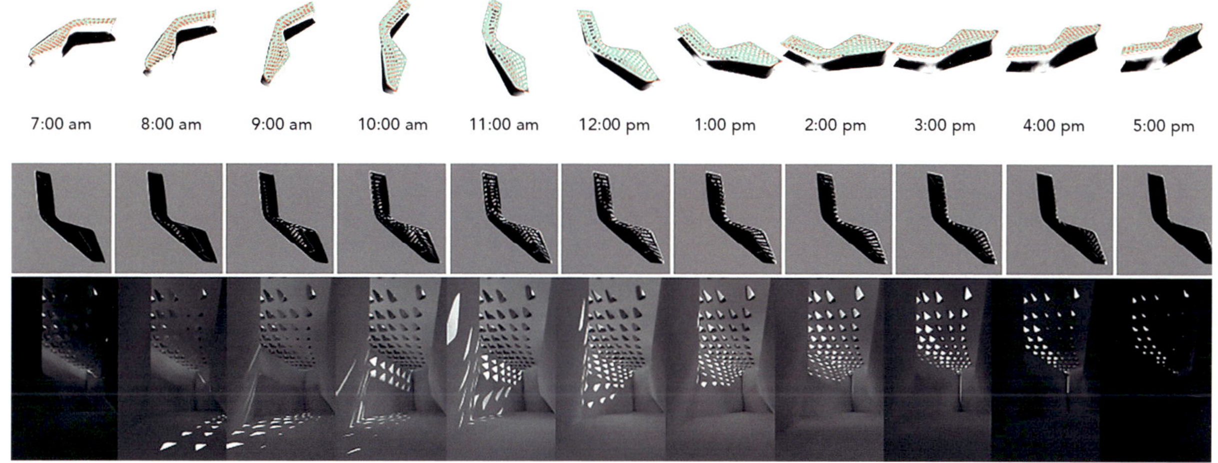

13

13 Zaha Hadid Architects with AKT II,
Guggenheim Hermitage Museum, Vilnius,
Lithuania, 2007.
Illumination analysis of the optimised roof
openings.

series of separate operations with distinct names. They provide a
vocabulary through which the user thinks about and communicates
the design process and its outcomes. This can be seen in the way
design students tend to talk about form using words that describe
a series of actions selected from a toolbar, rather than shapes or
construction techniques.

As digital media becomes ubiquitous, the design of interfaces
that facilitate the architect's access to underlying data becomes
an architectural problem in itself, which is perhaps too important
to be left to software engineers alone. Every custom-made
application encompasses a certain design paradigm that makes
assumptions and suggestions regarding priorities, values and
the distribution of roles that ultimately influence the designers'
intuition and the group's structure and communication.

The applications presented here operate at a granular level,
being project- or problem-specific. However, shifts and trends
can be observed in computer graphics, design software and
simulation applications that reflect technological changes
and emerging paradigms. One such instance is the transition
from the assembly paradigm (itself derived from mechanical
engineering applications) inherited from the Industrial Revolution
and embedded in most design software, to the biomedical
paradigm. In topology optimisation, we see a glimpse of such a
transition, where we can conceptualise the problem of structure
not as one of assembling appropriately dimensioned parts, but
rather one in which we distribute material properties in space.
Emerging fabrication and imaging technologies, metamaterials,
and simulation and computational tools, strongly suggest such
a paradigm shift in material culture. This is not because of any
romanticisation of nature that would lead to blind biomimicry, but
rather due to the increasingly fine scale of control over matter, and
an emphasis on the effect, performance and experience, rather
than the appearance, of material artefacts. New computational

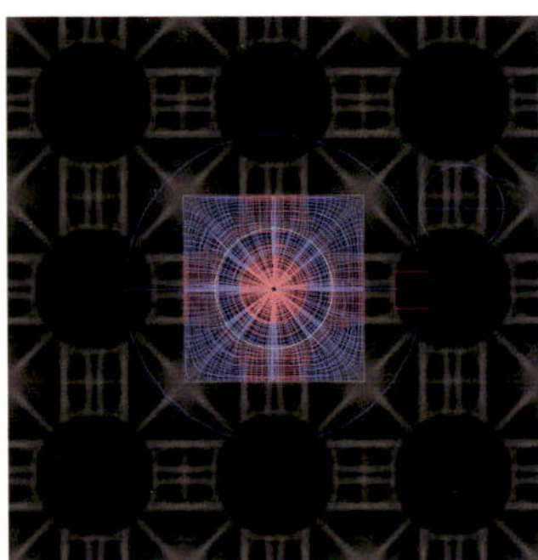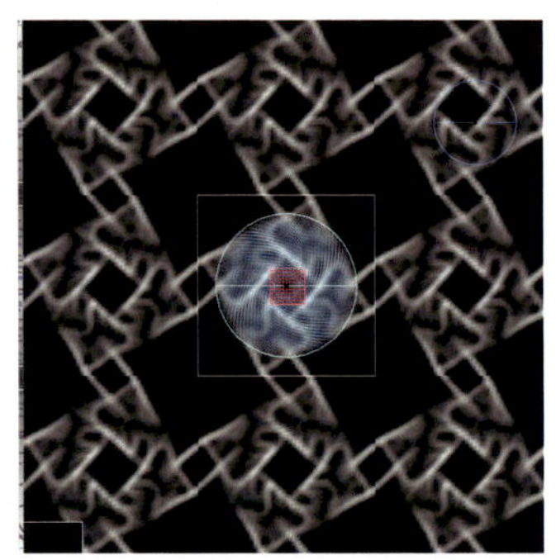

14

14 Topostruct, 2008.
Periodic structural patterns
using Topostruct. Topostruct
is developed based on
the theory of topology
optimisation.

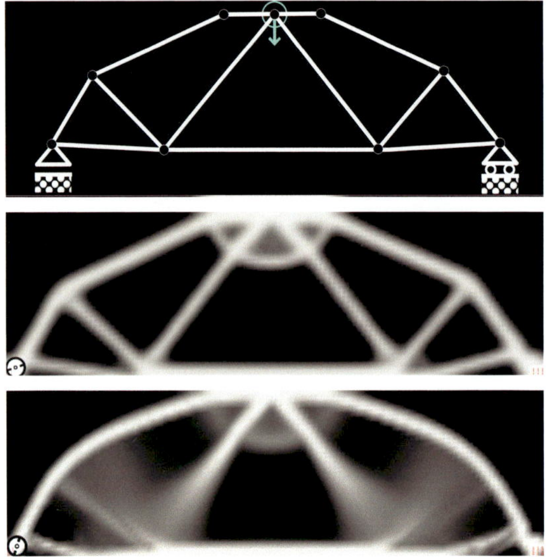

15

15 Topostruct, 2008.
Topostruct shifts the designer's
perspective of a truss being
an assembly of elements to
one composed of continuous
material distribution in space.

16 Topostruct, 2008.
Screen capture of Topostruct
illustrating the emergence of
natural forms.

17 Topostruct, 2008.
Topostruct provides multiple
visualisation schemes and an
understanding of notionally
optimal material distribution
in space.

16

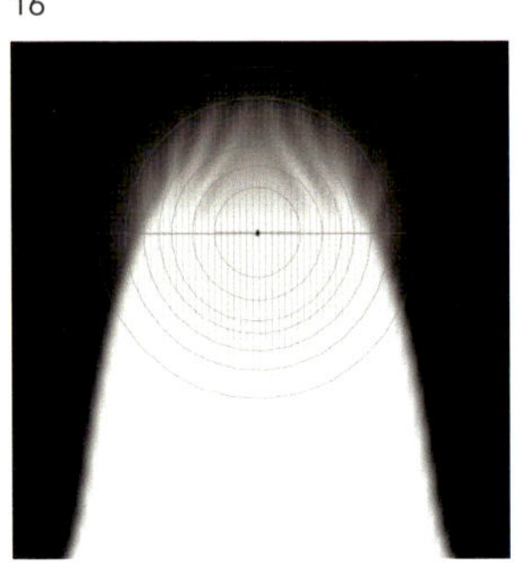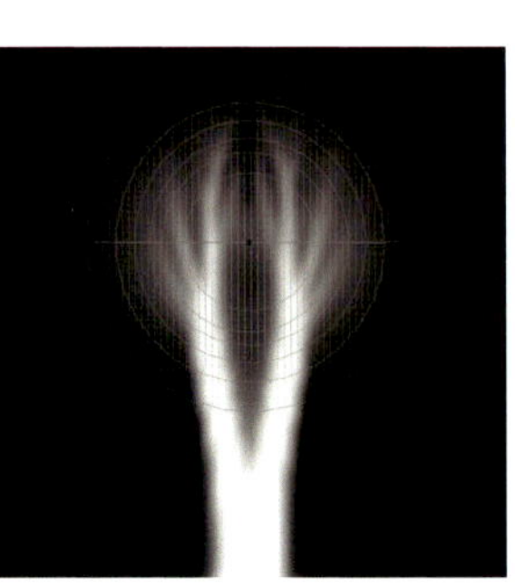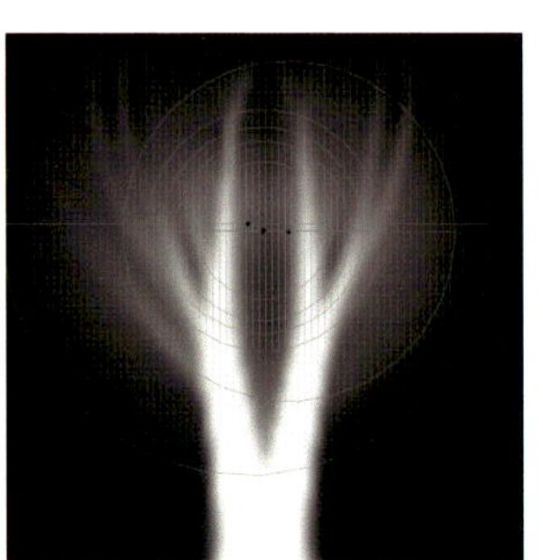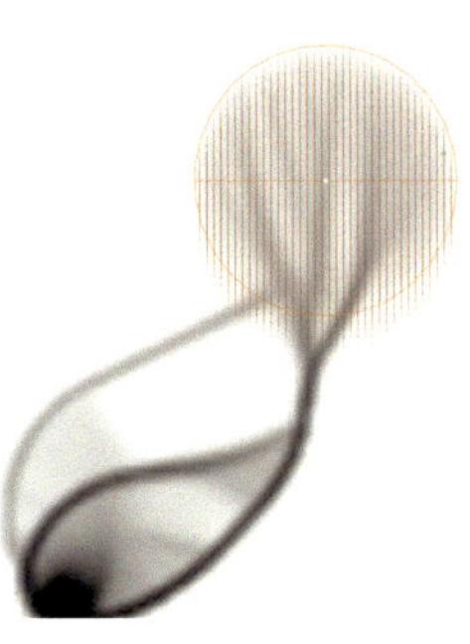

interfaces will be indispensable if architects are to engage with the possibilities opening up from design and material processes that span multiple scales.

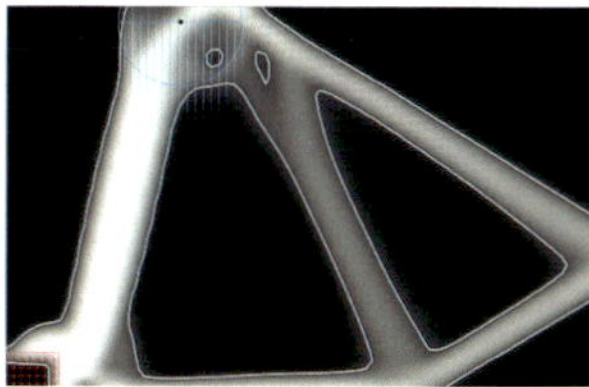
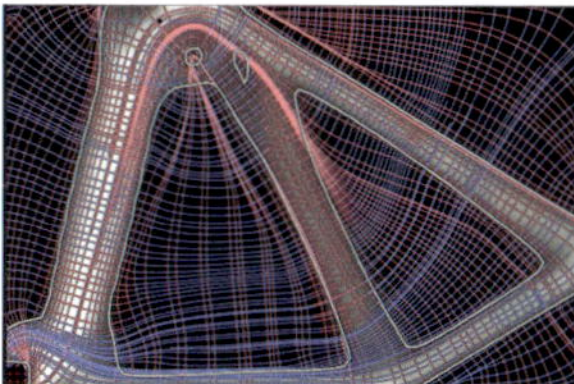
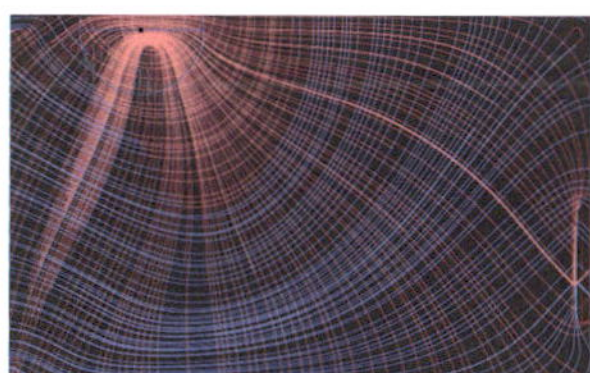

17

REFERENCES

1 In a 1973 essay entitled 'Structure, Construction, Tectonics', Eduard Sekler defined the tectonic as a certain expressivity arising from the statical resistance of constructional form in such a way that the resultant expression could not be accounted for in terms of structure and construction alone. E. Sekler, 'Structure, Construction, Tectonics', 1973, referenced by Kenneth Frampton in *Studies in Tectonic Culture*, MIT Press (Cambridge, MA), 2001.
2 Branden Hookway, *Interface,* MIT Press (Cambridge, MA), 2014.
3 Finite element analysis (FEA) is a computerised method for predicting how a structure reacts to real-world forces, vibration, heat, fluid flow and other physical effects. (Autodesk, http://www.autodesk.com/solutions/finite-element-analysis)
4 SOFiSTiK is a finite element analysis software which allows the modelling and analysis of complex doubly curved NURBS surfaces.
5 Nicolas Ray, Wan Chiu Li , Bruno Lévy, Alla Sheffer and Pierre Alliez, 'Periodic Global Parameterization', *ACM Transactions on Graphics,* 25 (4):1460–85, 2006.
6 Martin Philip Bendsøe & Ole Sigmund, *Topology Optimization: Theory, Methods and Applications*, Springer (Berlin, Heidelberg, New York), 2002.

TEXT

IMAGES

5 ABACUS AND SKETCH

ANDREW RUCK

The practice of engineering, like science or any progressive strain of thought and practice, is one that always stands on the shoulders of its forebears. The last 25 years have witnessed an extraordinary evolution in the way engineering, particularly structural engineering, is practised, thought about and discussed. The design of unprecedented structures, once within the reach and capability of just a few seminal 20th-century design practices, now appears to be within the grasp of many. This certainly might seem to be the case, but is it true?

1 AHMM with AKT II, 240 Blackfriars, London, UK, 2014.
The two distinct components of cladding and structure are visible during construction of this concrete and steel-framed office building.

2 Notre-Dame Cathedral, Reims, France, 13th century.
A medieval cathedral, 'Structure = Architecture'.

2

Technological advancements mean that computing power is now available on the high street for a modest price, so all engineering practices can equip themselves with highly sophisticated and powerful tools for design analysis, calculation, drawing and modelling. This provides an individual with the opportunity to draw and analyse structures of extraordinary feats of 20th-century engineering, such as the Sydney Opera House, in a matter of days; originally, its design took a huge team of exceptional engineering designers, researchers and computer programmers many months. What does that mean for the contemporary design engineer? Is he or she more able than before? Is practice so radically different from just forty years ago? Can anyone now design the extraordinary?

For the newly qualified graduate engineer, born and educated in an IT-enabled world, the current tools and language of practice are second nature, but for those who have been steeped in late-20th-century engineering practice, this revolution has been more palpable and uncomfortable. Many design engineering practices will have their own mix of Generation X and Generation Y participants, which can make for a lively and heady mixture of experience, design ability, drive, expectation, naivety and technical prowess. When these ingredients are harnessed, well-managed and directed in a collective endeavour, they can make for a formidable team capable of amazing things. Equally, when not orchestrated, they can result in a dysfunctional combination.

This chapter will scratch beneath the surface of some of the questions posed above, reflecting on the current nature of practice. It will look at the process of design and some of the tools used, while wondering whether current practice is really so different from that of its forebears.

3

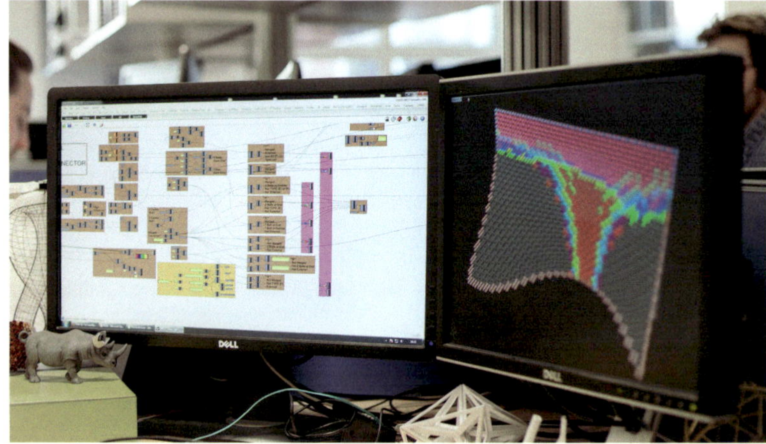

4

The role of the design engineer is largely concerned with 'designed' buildings and structures that have an inherent system by which they stand up and resist the forces to which they are subjected. Individual structures can be distinct (a definable component of the whole) as is, for example, the frame of an office building that is clad and concealed by a separate layer (Figure 1). In contrast, they can also be an integrated part of a holistic design, such as in medieval cathedrals where the visual manifestation of the whole building is also its structure (Figure 2). All such structures are preconceived, explored and tested prior to construction, and their performance anticipated to varying degrees. They are the product of a design process that is generally reductive and which converges towards an answer for each project. Beyond the single project, design engineering also exists as part of a long-term continuum of practice, with projects, areas of research and ongoing study accumulating into a body of work and thought.

In this context, the process of design is one of exploring ideas, with the principal aim of arriving at an appropriate solution. It involves the generation and definition of ideas, and their filtering and testing. It is a practice carried out by individuals or groups, involving debate, discussion, sometimes conflict, and the testing of an idea, either oneself or with peers. In the world of the design engineer, it also includes systematic testing using mathematics and calculations to give objective assessments of the validity and performance of an idea.

Sketches, drawings, diagrams and models are the tools of designers. They allow ideas, and aspects of those ideas, to be visualised, articulated and interrogated; they permit communication. The human brain simply does not have enough capacity to conduct the entire process of design in an imagined, virtual medium. Words, conversation and description are equally inadequate on their own; it is essential to create some sort of manifestation of a particular idea. The thoughts in the brain need to be embodied in some sort of physical medium so that further waves of iteration and testing can take place; something more than just the brain is needed.

Sketches have always been a mainstay of design, permitting rapid idea generation and concept development: ideas can be drawn then binned. And sketching, because it doesn't necessarily require the investment of hours and hours of work, is liberating and carefree too. It is also extremely beneficial in freeing the mind to explore the boundaries of a problem without the constraints of technical drawings, title blocks and CAD drawings. The latter, by giving the illusion of a degree of precision, can misguide or be positively misleading.

When sketching, there is something that happens between brain, hand and paper which helps to etch an idea into one's consciousness: some designers have said 'I sketch to see'. Sketches allow ideas to be recorded and then communicated to others, encouraging conversation and debate; often this discussion triggers further ideas and opportunities. Indeed, when asked about the origin of a really good idea, it is often difficult for designers to pinpoint it, as it often emerges from the interstitial spaces between designers during discussion and exploration.

Drawings are a similar staple for the engineering designer as a more formal means of recording and sharing ideas, articulating them more precisely and testing them, as well as ultimately providing precise detailed instructions for a builder to work to during construction. Historically these were produced by hand, by engineers and draughtsmen skilled at technical drawing; then, with the naissance of CAD, they were moved into an electronic medium.

Physical models are also powerful tools for the designer, permitting a solution or idea to be readily viewed from an infinite number of different perspectives. They allow an understanding to emerge that is not possible in a 2D medium, though they have been an underused tool in the engineering design office due to the time and investment needed in their creation. This has confined their use to the presentation of a solution at the end of the design stage. Architects, on the other hand, tend to use 'sketch-models'

3 Sketching design ideas.
Sketching has always been a mainstay of design. It permits rapid idea generation and concept development. Ideas can be drawn then binned. Because it does not necessarily require hours and hours of work, it is liberating and carefree. It frees the mind to explore the boundaries of a problem unconstrained by the formalities of technical drawings, title blocks and CAD drawings.

4 AKT II, design office, 2015, digital tools for visualisation and exploration.
A design engineer is using SOFiSTiK and Grasshopper to explore a piece of design.

5

5 Christopher Wren, St Paul's Cathedral,
London, 1675–1710.
Photograph of the dome of St Paul's. The
dome-shaped timber roof conceals the masonry
cone that supports the lantern and the inner
masonry dome that forms the ceiling.

6 Christopher Wren, St Paul's Cathedral,
London, 1675–1710.
Cross section of the dome of St Paul's
Cathedral.

more regularly as part of routine design exploration and idea generation stages.

Whether an abacus, slide rule, calculator – or, latterly, spreadsheet or software analysis package – is used, calculation is a skill the practising engineer needs and applies. Knowledge of the properties of materials, the mechanics of structures and the application of mathematics are used to assess the performance of a structure or element within. This draws on mankind's empirical evidence from past experience, successes and failures, while also extrapolating from the results of academic research and laboratory test data to inform the most credible outcome for a given structure's performance. Factors of safety are applied within this process to provide an acceptable and tolerable margin for error. In centuries gone by, this process leaned more heavily on past precedent, case study evidence and rules of thumb.

In among the tools and techniques used in design, there is generally a cross-fertilisation and evolution of knowledge, experience, ideas and approach, which develops with each successive project. While most buildings are one-offs, there is a degree of prototyping, testing, evolution and improvement that occurs as individuals and practices mature and develop their design skills, whether by refining design solutions and details, or by developing their approach to the practice of design itself.

CASE STUDIES
ST PAUL'S CATHEDRAL

The design of the dome of St Paul's Cathedral in London (Figures 5 and 6), part of one of the largest building projects of the 17th century, was carried out by Christopher Wren (known then as the Surveyor of the King's Works) and his team. They used physical models, drawings and sketches to define proposals and test them. Models were made of wood, including the Great Model of the Cathedral that was used to communicate their ideas and convince those commissioning the building, and there was a heavy reliance on the ability of highly skilled craftsmen to interpret the drawings and construct the building without the need for every single element to be defined. There was also a degree of design evolution during construction, through observation and experience.

At the time of the St Paul's commission, Wren was involved in the design of numerous other churches in London, which afforded him an opportunity, prior to the construction of St Paul's, to test out how to reconcile the geometry of a dome supported on eight piers. In the course of his work, Wren consulted and worked with the polymath Robert Hooke who assisted in the definition of the geometry for the masonry cone and dome, which form part of the cathedral's main roof structure. In their design process, Wren and Hooke sought to identify the pure geometry for the two structures. In modern parlance, they were undertaking 'form-finding', and successfully defining the geometry for a massively

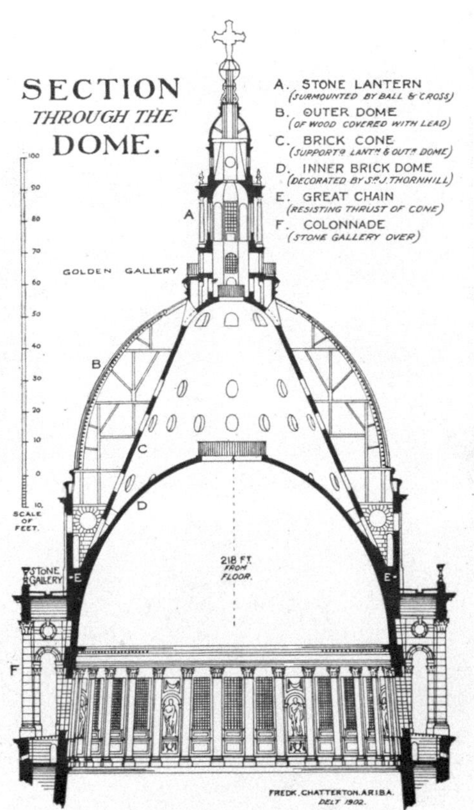

7

8

ambitious compression structure; this was an extraordinary feat of engineering and design by any standards, let alone in the 17th century, given the tools available. This is an early example of an architect–engineer collaboration, although the two characters involved are unlikely to have recognised it in those terms.[1]

MUNICH OLYMPIC STADIUM

In the mid-20th century, the German architect/engineer, Frei Otto, was studying and exploring lightweight and tensile structures. His magnum opus was the engineering of the hugely ambitious roof for the Munich Olympic Stadium in 1972 (Figure 7), working in collaboration with architect Günter Behnisch, who had won the design competition for the project. Otto was experimenting with natural phenomena, which included using soap bubbles as a proxy for the optimisation of surface geometry to inform his designs for lightweight tensile structures. He also used extensive physical scale models made from different materials to simulate the performance of the structures, and was a pioneer in the use of computers to assist with structural analysis.

His daring is all the more impressive, as he and his team were inventing the tools to facilitate their design work in parallel with actually doing the work. Designing something innovatory and extreme, while not knowing whether you are going to be able to discover the tools and techniques necessary to realise your ambition, would have been unnerving. This was truly pioneering design engineering work.

SYDNEY OPERA HOUSE

Jørn Utzon's masterpiece, Sydney Opera House (Figure 8), designed in collaboration with Ove Arup & Partners, was completed at a similar time to Otto's work. Its design and construction marked a seminal moment for engineering endeavour in the 20th century, remaining an extraordinary example of architectural engineering and collaboration, as well as technical and analytical prowess. The story of the project is well documented, but it is interesting to reflect upon the tools and techniques employed by Arup and Utzon.

Many published articles make reference to sketches, drawings and models as everyday tools in use. Computer usage was in its infancy so, similarly to Otto in Germany, the Arup London office pioneered software to analyse the geometry and structural performance of the concrete shells, and the tools it created were used to define the form of the iconic roof. Given the size and cost of computers at that time, this marked a huge company commitment to investment in technology.

The design of the roof was also tested on scale models in wind tunnels at the University of Southampton. These tests helped to establish the distribution of wind pressures around the roof in high winds. The results were then used in the design of the primary structure and the roof tiles and their fixings.

7 Frei Otto and Günter Behnisch, Olympic Stadium, Munich, Germany, 1972.
The design used a cable-net structure to support the roof finishes, tensioned against anchorages in the ground and a series of steel struts.

8 Jørn Utzon, the Sydney Opera House, Australia, 1973.
Physical model used to explain the form of the shells of the Sydney Opera House. Each of the shells is cut from the surface of a sphere.

9 Edward Cullinan Architects, the Weald and
Downland Museum Gridshell, West Sussex,
UK, 2002.
The internal timber lath structure exposed.

As Peter Murray stated in *The Saga of the Sydney Opera House*: '… the two men − and their teams − enjoyed a collaboration that was remarkable in its fruitfulness and, despite many traumas, was seen by most of those involved in the project as a high point of architect/engineer collaboration'.[2]

As well as being a high point in the use of technology to aid design, it is also remarkable for the final concept's simplicity, using the surface of a sphere to define the ribbed shells of the roof. This allowed repetition of form to simplify construction as well as to reduce the quantum and complexity of the engineering structural analysis.

DOWNLAND GRIDSHELL

Finally, the Downland Gridshell (Figures 9 and 10) in Sussex, England − designed by Edward Cullinan Architects with Buro Happold as engineer, and completed in 2002 − comprises a series of intersecting oak laths which are overlapped and woven to form a lattice. This lattice is then shaped to form a doubly curving surface. The project is a fine example of architect, engineer, constructor, craftsman and client all working together, striving towards a powerful collective vision. The construction sequence, which involved lowering the sides of the timber lattice to achieve its form, rather than the more conventional 'push up the middle', is a simple solution not born out of technology or analysis, but from basic engineering and construction know-how. Again, Buro Happold used customised software to map the structure. Mock-ups of joints and experiments with different materials were all part of the process and, like the Sydney Opera House, the project seems to have benefited hugely from a healthy dialogue and creative exploration of ideas and possibilities within the design team.

CONCLUSION

There are some common threads in all of these case studies. The designs have all been the product of a strong creative process; good chemistry between engineer and architect is prevalent, with dialogue and collaboration evident, and they all required the designers to create the tools necessary to carry out their design work. Sketching, drawing, models, structural analysis and material research were all key tools of the design engineer. The projects all required form-finding, and the structure and architectural expression are intrinsically linked. Ultimately, they were also the products of great teams and brilliant minds.

From these case studies it can be concluded that, while the tools of the design engineer have changed and evolved over the last 25 years, perhaps the practice of design engineering is not so different from that so successfully witnessed by the exceptional practitioners of the past. Those practitioners made their mark and contributed to the progression of design engineering, drawing inspiration from their predecessors.

The underlying process of design and design evaluation remains the same, and has not been supplanted by technology.

10 Edward Cullinan Architects, the Weald and Downland Museum Gridshell, West Sussex, UK, 2002.
External view of the building.

Professional practice has changed and the IT revolution has empowered contemporary design engineers – even emboldened them – to push the boundaries of design and construction further, possibly deluding them too at times. It has also served to promulgate and amplify the perception that the art of design engineering at its best is most evident in pure feats of modern structural engineering prowess, such as large domes, shells, long spans and tensile roofs. While these are among the more overt and expressive examples of a design engineer's skills, their fingerprints can make a similar, often unseen, contribution to more conventional buildings and structural types; great design engineering does not necessarily need to be arrogant or expressive. Moreover, it is a deep-rooted approach to the practice of design and engineering, with an uncompromising and insatiable appetite to seek out the best answer to a problem or challenge. It involves choosing the appropriate tools along the way; sometimes inventing them if necessary. It is a progressive and learning venture that draws from its own experiences and those of others, present and past.

Whatever the available tools and techniques, it is still a human process, practised by enquiring people; the tools are merely there to serve. Great chemistry between design team members and a healthy, challenging dialogue between parties is a necessary requirement.

The sketch, whether by pen and paper or through a piece of software, is still as relevant now as it has ever been. What is important is that the tool used continues to permit the mental connection that allows the designer to 'see' and engage with a creative design process. An appropriate visual manifestation that can quickly be used and cast aside is still essential. A similar logic applies to the tools used for calculation and the simulation of engineering structures. They too have evolved, and are exponentially more powerful than in the past, permitting thousands of iterations to be analysed and tested. This can be valuable, but when misused it can equally lead engineers to stop thinking for themselves, thus reducing the need for some of the critical evaluation that could ordinarily prevent the need for certain iterations to be considered.

While the high street might sell the necessary technological tools to anyone with the means, there is still the fundamental need for a particular individual to operate, use them and think critically. Armed with the tools, it is the user (individual or collective) that makes the difference between engineering nonsense and something of value.

AKT II operates within the context of trying to harness and exploit the power of modern technology, mindful that the right tools need to be used for the right job. It is prepared to make them too, if necessary. Its staff currently encompasses a broad range of ages, successfully blending the different skills of generations X and Y. Also, the full armoury of tools, past and present, is available to suit the user, and the practice is capable of designing overt expressions of engineering prowess, but is equally adept when a subtle, nuanced engineering answer is demanded in more everyday structures. It inculcates a progressive technology-led environment within the office, which permits past experience from its own projects to be readily available so that it can be brought to bear on the evolving body of work. It is content to draw inspiration from peers, past and present. AKT II exists as another punctuation point on the timeline of design engineering protagonists.

REFERENCES
1 Derek Keene, Arthur Burns and Andrew Saint, *St Paul's: The Cathedral Church of London 604–2004*, Yale University Press (New Haven and London), 2004.
2 Peter Murray, *The Saga of the Sydney Opera House*, Spon Press (London), 2004.

6 DIGITAL DEXTERITY DJORDJE STOJANOVIC

In the modern day, design engineering is increasingly discussed as a new model for working in the construction industry. The emphasis is placed on the collaborative nature of the model and, in particular, on the partnership between architects and engineers. It seems only logical that collaboration is the only way forward when undertaking complex tasks like building projects, but it could equally be worth considering that this developing proximity between the two disciplines, fuelled by advancements in digital technology, produces certain ingenuities and dexterities within each discipline. For engineers, their input is more integral to the early, conceptual stages of the building project. This suggests, from the architect's point of view, that not only should structural principles and material characteristics be taken on board early in the design process, but that such principles could also be transposed and applied to fundamental thinking about spatial organisation, thus influencing social and cultural aspects of the built environment.

DESIGN STUDIO CURRICULUM AND MATERIAL-BASED STUDIES
At the University of Belgrade in 2008, several topics relevant to both architectural design and structural engineering were central to the established educational agenda, representing an intersection between architectural practice, education and research. At that time, these topics were formulated as the following set of research areas and teaching topics within the design studio curriculum:

- Applying computational logic to thinking about spatial organisation and construction through both analogue and digital means.
- Developing a better understanding of material properties (ie, elasticity) with the use of digital tools.
- Learning through hands-on building experiments and making prototypical models.
- Acquiring transferable knowledge, rather than technical skills.

In parallel with producing building projects, students on various design studio courses were also given abstract exercises in the form of material-based studies[1] conducted as a sequence of experiments. Over the years, a number of prototypical models

1

1 Djordje Stojanovic,
prototypical model,
'Inconsistencies v.03',
Belgrade, Serbia, 2012.
The model was produced as
an outcome of the design
studio curriculum at the
University of Belgrade, Faculty
of Architecture. It was created
jointly by students and staff.

were produced to test the practical and theoretical dimensions of the design approach that employs rule-based logic and elastic material performance to achieve highly versatile spatial organisation. Each semester, students were asked to work collectively to build large structures, such as the construction of an assembly to fill an entire room: 12 kg of yellow rubber bands measuring 70 mm in length and 5 mm in width, and approximately 8,000 metal clips for joints, were involved as construction components. The resulting structure occupied the room with a footprint of 50 m^2 and a height of 3.5 m (Figure 1).

It is important to note that students were not asked to create their own individual models, but to participate in the creation of a single structure based on established design protocols. No drawings or computer models were made prior to the construction process; instead, students were given a set of rules and verbal instructions formulated from the knowledge gathered in previous experiments. Construction started simultaneously from five points, from which a number of 'tentacles' were established in relation to the structural considerations of the most suitable supporting points within the given environment. From there the structure grew through the addition of new tentacles at the mid-point of existing strands. As anticipated, after a number of recursive steps, the initial rule-based growth process became less apparent and had to give way to a new logic related to the elastic material behaviour or the inherent property of the employed building material.

Although the work was produced by students of architecture, the result of this experiment was not a building, nor was it presented as an example of what they should do in future practice. It is also worth noting that structural reasoning played a certain part in the construction process, but bore little relevance to the end result. What is the significance of such an approach in architectural education? And if this really is about acquiring transferable knowledge rather than the development of technical skills, how can future architects benefit from such experimentation?

INTERPRETATION AND DEVELOPMENT OF STRUCTURAL MODELS IN ARCHITECTURAL EDUCATION

There are various ways of looking at the relevance of this experiment and the resulting structure. First, the exercise is about developing manual dexterity and problem-solving skills that could be applied to many different tasks during the process of designing and constructing buildings. Second, the resulting structure could be observed from an organisational point of view and so is useful for an understanding of the built environment.

For Branko Kolarevic,[2] for example, a prominent characteristic of the structure is the distinction between the initial and the emergent set of rules employed throughout the construction process. In this case, such emergent rules are directly related to the material performance: through the effect of elasticity the entire physical model acquires an instantaneous ability to recalculate itself according to any amendment or addition of a new component. Throughout the growth process, the overall stability of the structure is reliant on a multitude of local conditions and the ability of identical modular components to react to tension forces and continually adapt to evolving structural circumstances.

Another interesting interpretation of the model is given by Mark Wigley[3] in the context of a broader conversation on the relevance of network organisations in architecture. When presented with images of the structure, Wigley was able to point out the resilience of the system, which he then recognised as the enabler of the curious spatial condition defined by the lack of distinction between the interior and the exterior of the created structure.

The structure could also be viewed as an environment created between elastic lines, as well as an object with its own structural logic. There are segments with differing densities within the structure, and a closer look reveals that different zones of the model have individual properties. The majority of segments with higher densities of elastic lines resemble objects with their own identities and boundaries, while other segments positioned closer to the existing walls reveal features of the environment, allowing visitors to walk through them.[4]

All the identified characteristics of the model – the ability to recalculate itself, the lack of boundaries between interior and exterior, and the seamless blending of individual segments – are related to structural performance, but could also be applied to thinking about how buildings and cities are conceived and used.

Other experiments, conducted in the same or similar format, include a large structure where more emphasis was placed on the building process and hands-on ways of learning. Instead of rubber bands, rolls of elastomer-based strips were used, and in the place of metal clips there were purpose-designed joints made of two laser-cut steel plates with two plastic ties to hold them together. The shortest span between two ends of the structure was 13 m and its height reached just over 5 m (Figure 2). Preparation for the construction of this model led to the development of custom computational tools that enact simulations of elastic material behaviour throughout the process of geometric modelling.[5]

In other scenarios, elastic material properties and rule-based construction techniques were employed to enable interaction with the built form. The prototype constructed at the University of Tehran, as a part of the Architectural Association's Visiting Programme, reached the height of 11 m. Similarly to the first two examples, this structure was designed to respond to externally applied forces by changing its geometric configuration and then resuming its initial state after the action, but this time it was done in relation to the force imposed by the weight of a person. To everyone's amusement, on the final day of the workshop, visitors and fellow students were invited to test the model by swaying in it with the amplitude of 3 m (Figure 3).

CONCLUSIONS

Although the driving force behind each of the three showcased structures is based on the knowledge transposed from contemporary design engineering practice into architectural education, such experimentation is relevant for architects, and not only in the learning of structural principles and material properties. Indeed, such prototypical models have been produced to test the practical and theoretical dimensions of the design approach, employing elastic material performance to achieve a highly versatile spatial organisation.[6] This approach has introduced distinct workflows in which the architect assumes only an indirect control of the model, allowing for a more open negotiation between material and structural performance in the design process. This unfolded as a series of feedback loops in which material performance, intuitive decision-making and computational tools were all combined.

With the expansion of design techniques and technologies, architects may have acquired an appetite for more complex and

2 Djordje Stojanovic, prototypical model, 'Inconsistencies v.04', Belgrade, Serbia, 2011. The model was produced as an outcome of the design studio curriculum at the University of Belgrade, Faculty of Architecture. It was created jointly by students and staff.

2

versatile forms. In response, educational settings aim to replace
post-rationalisation in architectural design, whereby formal thinking
is separated from material and structural reasoning. In structural
engineering, increased involvement with digital technologies
enables better understanding of material and structural properties,
and a similar kind of transformation is taking place in architecture.
New tools for analysis, simulation and representation also enable
better understanding of how buildings can be constructed,
as well as how built space is used and how cities grow. Some
digital technologies and design techniques originate in
structural engineering and other disciplines, but are increasingly
finding applications in architecture. It is precisely this exchange
between disciplines that produces a form of dexterity and new
developments in knowledge.

REFERENCES

1 The initial version of the text documenting the
development of the sequence of experiments titled
'Inconsistencies' was first presented at Delft University of
Technology, eCAADe conference, 18–20 September 2013.

3

3 Djordje Stojanovic, prototypical model, 'Inconsistencies v.02', Tehran, Iran, 2011. The model was produced as an outcome of the design workshop, organised by the Architectural Association, Visiting Programme. It was created jointly by students and staff.

2 Branko Kolarevic, 'Performance in Architecture', personal communication, 4 December 2012.

3 Mark Wigley, 'Interview with Djordje Stojanovic', in: Vladan Dokic and Petar Bojanic (eds), *The Specter of Jacques Derrida*, University of Belgrade, Faculty of Architecture (Belgrade), 2013, pp 34–49.

4. Djordje Stojanovic and Milutin Cerovic, 'The Cloud Model', in: Anastasios Tellios (ed), *Agile Design: Advanced Architectural Cultures*, CND Publications (Thessaloniki), 2013.

5 This software is now available as a plug-in for the Rhinoceros platform, under the brand name 'Spider'.

6 At the turn of the 21st century, two North America-based writers presented stimulating visions of plausible spatial organisations based on knowledgeable overviews of historic precedents in art and architecture. The first one was Stan Allen (1997) who depicted the Field Conditions as bottom-up phenomena, defined not by overarching geometrical schemes but by intricate local connections. In 2001 Mark Wigley described the Network Conditions as an effect that cannot be designed, something that does not have an interior or exterior, a system of interlocking elements with many similarities to biological organisms. Instantly after their publication, both essays became an integral part of a great many agendas in architectural education and research. The presented design studio curriculum establishes a connection between the two ideas – briefly described above and observed here as two theoretical models – and material-based studies in architectural design, conducted as a sequence of experiments and resulting in the series of prototypical models.

TEXT
© 2016 John Wiley & Sons Ltd

IMAGES
Figure 1 © Ana Kostic. The prototypical model was produced by: teaching assistant Milutin Cerovic, students: Milica Tasic, Ivana Radovic, Katarina Mercep, Marija Pop-Mitic, Danka Sijerkovic, Jovan Pucarevic, Dea Kolacek, Milos Simic, Emilija Zlatkovic, Milan Katic, Dusan Tasic, Sonja Elakovic, Ana Todosijevic and Marko Vukajlovic.
Figure 2 © Ana Kostic. The prototypical model was produced by: teaching assistant Milutin Cerovic, students: Milan Katic, Milica Tasic, Nikola Milanovic, Ivana Radovic, Katarina Mercep, Marija Pop-Mitic, Danka Sijerkovic, Jovan Pucarevic, Dea Kolacek, Milos Simic, Emilija Zlatkovic, Dusan Tasic, Ana Todosijevic, Marko Vukajlovic and Nevena Bjelakovic.
Figure 3 © Djordje Stojanovic. The prototypical model was produced by: tutors Djordje Stojanovic and Maryam Pousti, students: Amir Reza Esfahbodi, Abolhassan Karimi, Imman Shameli and Mohammad Habibi Savadkuhi

1

2

1 Sebastian Partowidjojo, private residence,
Surabaya, Indonesia, 2015.
Close-up of a brick wall in Surabaya, Indonesia.

2 Sebastian Partowidjojo, private residence,
Surabaya, Indonesia, 2015.
A bricklayer constructing a wall for a private
residence in Surabaya, Indonesia.

7 DIGITAL VERNACULAR

JEROEN JANSSEN AND
ADIAM SERTZU

Architecture has continually been informed through its representation, production and construction; however, with the rapid increase of emerging technologies, it has now moved beyond the limits that we had imagined possible. Nowadays, the availability of digital software, and the ease with which it facilitates design, makes designing without it unthinkable. The last few decades have seen the emergence of a new style of digital architecture, driven by growing hardware capabilities and an array of computational design software that can generate any curved geometry. Several projects have been built with the help of bespoke fabrication methods and held up by over-designed structures, merely providing the shape for the designed geometries. Most examples seem to have been driven by the extensive possibilities and embedded features of the software, which allowed designers to unleash their imagination and become detached from physical reality, rather than be driven by performative requirements. It is in such instances that the link between representation, production and construction seems to be lost in translation.

In 'Translations from Drawing to Building',[1] Robin Evans discusses the dissociation between the two that architects face. Digital production, like drawing, is a generative medium that comes with its own set of restraints and possibilities. As with all tools of production, the very techniques that allow these explorations have their own constraints, each with specific working methodologies. Most digital architecture does not have the same 'quality' as a hand drawing; it does not carry the same amount of data or information that is usually embedded in an old-fashioned sketch, and few architects are proficient enough to really use the digital tools as a sketch paper.

DIGITAL FABRICATION

The global architecture emerging from this method of digital design has proved impractical: while fabrication techniques have not been able to catch up with these novel modes of design, architectural design in turn has not merged with fabrication methods in order to make the projects any more buildable. Furthermore, most of these designs are put together without consideration of the local context or technological advances, and

3 Light horn.
Concrete facade mock-up
produced during the EBE
workshop at the EiABC in
Addis Ababa, Ethiopia.

4 A craftsman preparing
clay moulds for cooking
stoves from recycled metal
at Menalesh Tera in Addis
Ababa, Ethiopia.

4

therefore remain disconnected from their immediate surroundings.
This has resulted in an unworkable status quo, since the growing
pressure created by the global scale of urbanisation demands an
immediate answer to this architectural crisis. At the same time,
advances in digital fabrication have instigated a design revolution,
generating a wealth of invention and innovation that is sparking
the imagination of a new generation of designers. A network
of activities has grown around this method of fabrication in a
relatively short period of time.

Fab labs and other incubators are changing the cultural landscape
of many places through the design ingenuity emerging from digital
fabrication and the material practices they have created. There
are many fab labs in developing countries that want to harness
these new technologies, and rightfully so; making should not be
restricted to the abstract digital realm, but should continue to
become a force for innovation and investigation in architecture,
just as it has done for thousands of years. Building should be seen
not just as the implementation of represented conceptions, but
rather as a process by which one explores and delivers for, and
within, a specific place.

5 A craftsman proudly showing off his workmanship at Menalesh Tera in Addis Ababa, Ethiopia.

WHAT CAN BE LEARNT FROM VERNACULAR ARCHITECTURE?

Rather than being a style, it is completely driven by performance. It could be argued that, with its self-development and self-criticism over time, vernacular architecture is the supreme evolutionary survivor for a specific programmatic question, local context, community and technique; it is a delicate equilibrium between form and function. This architecture has a quality in sustainability and spatial experience that is usually not driven by intended design, but instead through experience and craftsmanship passed on from generation to generation. Mainly for this reason, this mode of construction occurs at a small, local scale. To address the current crisis of urbanisation and to learn from this vernacular discourse, a paradigm shift is essential. Fortunately, design engineers nowadays have the opportunity of using novel digital design tools to establish this shift. The qualities of vernacular architecture can be examined to determine why they are so successful; they can be embedded within models, and information about digital fabrication can help to generate intelligent and informed design systems. It follows that a new balance for the self-organising and delicate equilibrium between form and function will have to be found, and, this time, data and digital fabrication have to be incorporated into the equation.

Perhaps the most important aspect to embed in the design process, especially when looking at building locally, is the local climate of a site. The digital design tools currently available allow for the incorporation of climatic inputs in the design process as an inherent attribute, creating a responsive design model. Furthermore, the way different buildings are used can be monitored during their lifetime and these rich datasets utilised to simulate future behaviour, allowing for designs to be adjusted accordingly. These are just a few of the opportunities design engineers currently have to hand, presenting a vast array of possibilities.

ADDIS ABABA EXPERIMENT ON BUILDING ENVELOPES

One striking example of the paradox of using glass facades in hot climates is Addis Ababa, a city with a compelling history and one of the fastest growing urban areas in the world. Now the 10th largest city in Africa, Addis Ababa is currently rebuilding its urban core and enlarging its periphery, causing a radical transformation to its skyline as office blocks, malls and low-cost apartments bloom throughout the city. This rapid urbanisation creates a complex and dynamic system in which it is difficult clearly to predict future construction and typological trends. However, it seems apparent that, as in many other developing nations, the glazed facade system has already emerged as the preferred building envelope.

It is because of this prototypical context that Addis Ababa was chosen as the focus for a course called 'Experiment on Building Envelopes'; a collaboration that brought together artists, architects and engineers to look critically at the efficiency and suitability of glass facades within the particular climatic conditions of sub-Saharan Africa. This congress sought to challenge the existing approaches and explore alternatives: is it possible to design

6

6 Stacks of recycled and reshaped steel pans for sale at Menalesh Tera in Addis Ababa, Ethiopia.

8

7 Project Echo.
Steel facade mock-up
produced during the EBE
workshop at the EiABC in
Addis Ababa, Ethiopia.

8 Metalworker welding
connection points for the
individual discs in a steel
facade mock-up during the
EBE workshop at the EiABC in
Addis Ababa, Ethiopia.

facades that reflect local material and culture within this context? Can we move away from imported facade construction techniques and off-the-shelf component systems? How can advanced design technologies be made more relevant when engaging with local and low-tech skills and materials? And lastly, how can design approaches be adapted when considering existing inefficient building skins?

This discussion was influenced by Bernard Rudofsky's *Architecture without Architects*,[2] in which he rejected the idea of standardised dwelling concepts, and instead promoted the idea that the built environment should reflect the history, culture and climate of the immediate surroundings. For Rudofsky, architecture is more than just the exposition of aesthetic styles; it is the framework for an intelligent way of being. These schemata find expression and realisation in the built environment through the confluence of many factors, principally material availability, manual fabrication skill and access to construction tools. The Experiment on Building Envelopes course therefore sought to renew this concept by considering how access to, and the understanding of, contemporary digital tools can allow this relationship to evolve. What opportunities emerge when local materials and craftsmanship are combined with technologically advanced digital software? Only through a higher understanding of where local material and culture and new digital tools align can the application of standard materials be reinvented and a new vernacular emerge. As stated in Kenneth

9 Completed steel frame and leather panels for a steel and leather facade mock-up during the EBE workshop at the EiABC in Addis Ababa, Ethiopia.

10 Processed leather hide at Menalesh Tera in Addis Ababa, Ethiopia.

10

Frampton's '10 Points on an Architecture of Regionalism',[3] this is not a vernacular in the sense of historical vernacular styles, but instead a contextually sensitive and progressive design approach that broadens the existing Ethiopian architectural discourse.

The brief for the course was to design an environmentally responsive facade system; within this framework, the first information that went into the digital design was driven by geometry. Facade concepts were broken down into sets of geometrically discrete components that responded individually to various inputs. These parametric components could align and resize in response to varied, even conflicting, requirements of the input datasets. This resulted in a digital model containing a set of simple and robust components that were fed with multiple inputs. While these inputs covered only one aspect of the design at a time and were easy to describe as singular rules, in concert they generated a highly intelligent composition.

This model was followed by a discussion on the technical delivery of the different facade concepts. Full-scale facade mock-ups were developed in parallel with each digital design, allowing the students to test them physically and gain a closer understanding of the actual construction process of facades, with the restrictions this can bring to design intent. During this phase, the students

received technical and structural engineering guidance specific to each design, as well as input from local craftsmen relating to construction. Structural sizes were determined for elements, and a large number of tests were conducted to find the appropriate methods and materials for final prototypes. For example, the concrete projects went through an extensive set of test pours to find the correct mix of concrete and the appropriate material and methods of construction for formwork. Connection details were designed, engineered and tested in the workshop with the help of local craftsmen. This knowledge was then fed back into the digital models and so informed the final designs before construction.

CONSTRUCTION AND INDUSTRY

Inventive methods have emerged from project-specific applications developed by a handful of designers and fabricators. By integrating the design, analysis, manufacture and assembly of buildings around digital technologies, the inventiveness of architects, engineers and builders now has a part in restructuring the very process of construction. One example could be taken from the timber construction industry in Switzerland and the south of Germany. Traditionally, timber construction was driven by carpenters and craftsmen with local small-scale workshops, working on a manual basis and serving only the local area. Over time they developed and reinvented their profession, and they are currently world leaders and experts in complex timber construction: no geometry is too complex, and digital fabrication is at the heart of their day-to-day work.

But methods utilising mass production cannot be simply replicated in the construction world. Many aspects influence the architectural design of a building: clients with different questions, occupants with varying behaviour, communities with local habits, programmatic variations, climatic differences between distant regions, etc. These demands require a specific response in every new design.

The ability to mass-produce digitally irregular building components within the same facility introduced the notion of 'mass customisation' to building design and production. Mass customisation is sometimes referred to as systematic customisation and can be defined as the mass production of individually customised goods and services. It offers a tremendous increase in variety and customisation, without the extra costs. As architects increasingly find themselves working across disciplines of architecture, material science and computer-aided manufacturing, the historic relationship between architecture and its means of production is further challenged by the emerging digitally driven processes of design fabrication and construction.

Building components designed within this realm of mass customisation are all digitally controlled. This enables design engineers to embed the possibilities and constraints of the production process directly into the early design models and feed into the newly defined equilibrium of form, function, data

11 Leather leaf.
Leather and steel facade mock-up produced during the EBE workshop at the EiABC in Addis Ababa, Ethiopia.

and digital fabrication. The management and installation of these components on the building site should also be considered and aligned within this equation, eventually completing the circle of design to production.

NEW MATERIALITY

New digital processes of conception and production have slowly permeated building design and production. The subtle variation of a system of elements, the transformation of recognisable materials and the visceral response to viewing the result of intensive material accumulation have all been digitally refined, giving rise to a new architectural language. Architects have to couple form with method along these lines, and revisit tectonic systems as a means to produce material effect. They seek to elevate standard building materials perceptually through non-standard fabrication processes; in an approach that recognises what Michael Speaks has termed 'design intelligence', 'making becomes knowledge or intelligence creation'.[4] In this way of thinking and doing, design and fabrication (and prototype and final design) become blurred, interactive and part of non-linear means of innovation. Many architects are committed to employing the fluid potentials of technology to inform the design process and gear the evolution of their designs, while their experimentation is remarkable for being on a one-to-one scale.

WHERE DO WE GO FROM HERE?

Looking back at the successful results of the Experiment on Building Envelopes course and the goals set out at the start, it can be seen how it made a small but valuable contribution to the development of a new 'digital vernacular' method of design: a method that does not view 'contemporary' as a mere 'copy-and-paste' approach, but looks back at the enormously rich local history of design and construction; a method that looks at interventions driven by well-considered local climatic responsive concepts, utilising all the contemporary design methods and tools available today. The workshop sparked new insights among the participants and furthered local discussions on Ethiopian architecture, as well as contributing to a broader international discussion on how to bring this vernacular architecture back to a well-deserved position of importance within the design profession.

There is still a long way to go: history shows that the construction industry is a slow-moving animal, afraid of change. However, the urgency of a rapidly urbanising world demands immediate action, and all the ingredients to create this necessary paradigm shift are at hand. The Experiment on Building Envelopes course – along with many other positive initiatives around the globe with a new digital approach to embedding design into the local context – will enable us to deal with all the exciting opportunities and challenges ahead.

REFERENCES
1 Robin Evans, 'Translations from Drawing to Building', in: *Translations from Drawing to Building and Other Essays*, MIT Press (Cambridge, MA), 1997.
2 Bernard Rudofsky, *Architecture without Architects: A Short Introduction to Non-Pedigreed Architecture*, University of New Mexico Press (New Mexico), 1987.
3 Kenneth Frampton, 'Towards a Critical Regionalism: Six Points for an Architecture of Resistance', in: *The Anti-Aesthetic, Essays on Postmodern Culture*, Bay Press (Seattle, WA), 1983.
4 Michael Speaks, *Versioning: Evolutionary Techniques in Architecture*, edited by SHoP / Sharples Holden Pasquarelli, Wiley Academy (Chichester), 2002.

TEXT

IMAGES

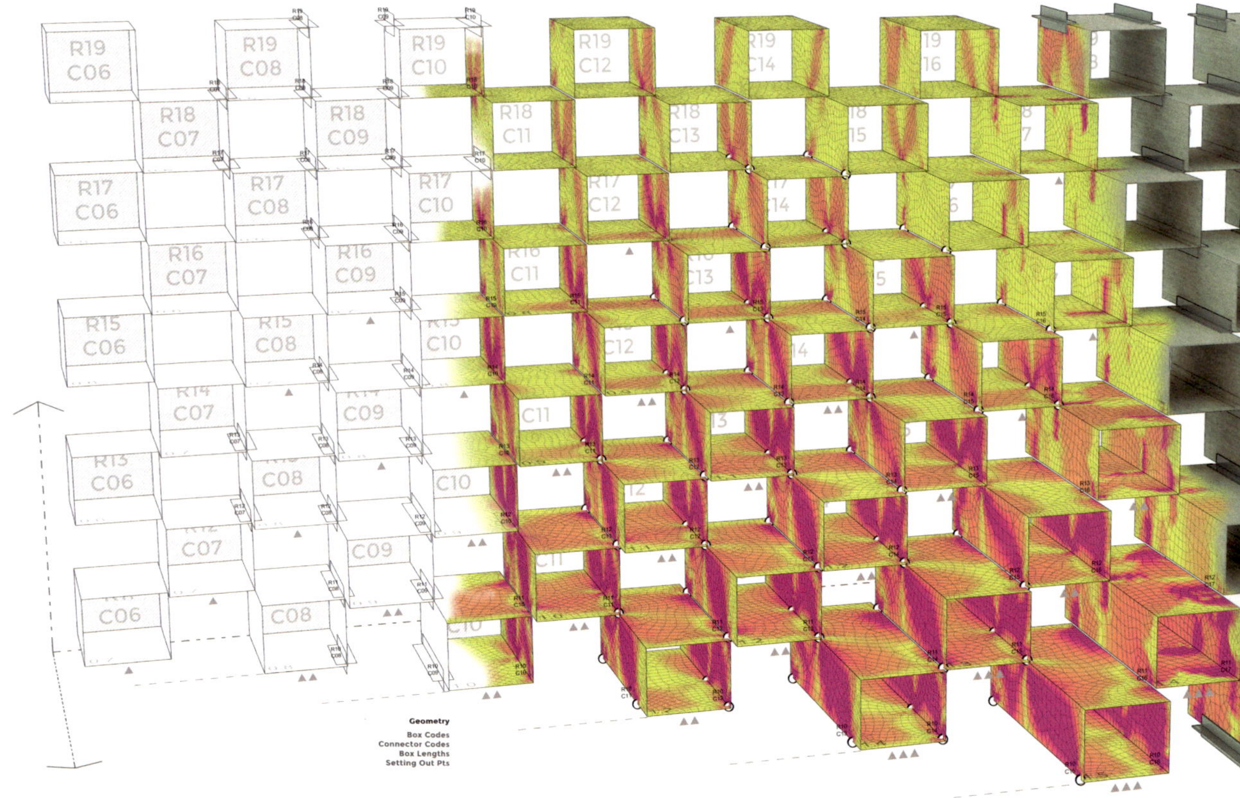

Bjarke Ingels Group (BIG) with
AKT II and AECOM, London,
UK, 2016.
Curving fibreglass block
twin wall of the Serpentine
Pavillion.

PART 2:
HEFT, ONTOLOGY AND HORIZON
DANIEL BOSIA

This section discusses a first-principles approach to some of the overlapping areas that form the domain of the design engineer's work today.

Attention is drawn to the way parts come together in space to form a structured whole, characterised by a network of effective load-bearing paths. With the post-Euclidean analytical approach of modern mathematics to the structure of geometry, the importance of topology is discussed through simple prototypical examples, as a means of organising matter and creating form and space. There is the beginning of a discussion relating to the importance of advances in material science and fabrication technology in the work of today's design engineers and the construction industry as a whole. Examples of some new materials with enhanced structural properties are presented in contrast to those that display multi-parametric performance. With the multidisciplinary engagement of structural and environmental engineering, the fields of structural fluid and thermal dynamics overlap and interact, creating new opportunities in the design of the building envelope and the public realm while showing the need to cross traditional boundaries of disciplines.

Following this, three emerging typologies are selected for closer examination to draw out the new ways of advancing them. The first typology consists of complex frame structures through to continuous monocoques and stress skins, where envelope and structure create a multi-parametric building envelope. The evolution of structural shells is then discussed, illustrating how these have evolved from pure compressive forms form-found under their self-weight, to intricate structures that can respond to more complex load patterns and holistic sets for architectural and environmental criteria. Finally, a look at the evolution of tensile structures, from simple catenaries to complex pre-stressed anticlastic and air-supported structures, puts this lightweight typology within the context of broader applications.

To conclude, the importance of the interoperability of bespoke computational tools in the collaborative design of buildings today is discussed. Real-time interactions between computer models and analysis are the new sketch tools of the architect and the design engineer, merging the creative synthetic process of intuition with that of forensic analysis.

8 GEOMETRY AND ORGANISATION DANIEL BOSIA

In Euclidean terms, geometry is the synthetic, intuitive mathematical description of objects in space, without a clear distinction between physical and geometric space. With Descartes's introduction of a coordinate system, Euler's and Gauss's studies of the intrinsic structure of geometric objects, and modern mathematics – with its strong links to general relativity – geometry has turned progressively more towards the analytical study of the topological relations between geometric entities such as points, lines, surfaces, solids and their higher dimensional equivalents.

1

2

Establishing continuity through the creation of physical load-paths is at the core of a structural design engineer's work; proportions and materiality usually follow. Topology governs the networks and organisations that underlie any structure and, as a consequence, have an impact on architectural space and form. Topology is the study of the way in which constituent parts are interrelated or arranged, their hierarchies and their connectivity. To create a structural network is to establish its interior logic, the map of its continuity in space; in other words, the organisational structure. Topological continuity can be achieved through a number of different methods of addition and subtraction. Stacking and branching are typically additive methods; erosion and subdivision are examples of subtraction.

In simple prototypical projects, these principles can easily be tested. For example, Canstruction – where teams of designers are asked to produce installations using simple modular food cans – is an opportunity for engineers quickly to investigate the creation of structures through vertical stacking, with continuous interlocking load-paths that consist exclusively of the self-weight of the cans shifting in space through key connection points. Here, the topological organisation of the can stacks and their connections at each level determine their strength and stability, their form and their appearance. In other words, where connections are established, load-paths can be formed through to the ground, and where separation occurs, space emerges. The geometric organisation of this structure consists of the subdivision in plan of three columns in clusters of three and sub-clusters of seven. Columns, clusters and sub-clusters are rotated about vertical axes in alternating clockwise and counter-clockwise directions to form a complex series of spirals within spirals (Figures 1 and 2).

Events like the London Design Festival offer another opportunity to prototype new geometric/structural concepts. At Testbed 1 in Battersea, engineers created a piece they named S-String (Figures 3 to 5), the prototype for an adaptive branching structure of modular units. Based on the use of a truncated triangular prism, connected to its neighbouring units at its square side-faces, it can be reconfigured geometrically by rotating each component by 0°–90°–180°–270° around the centre of each square face. Inherent within the geometry of the three square faces of the truncated prism is the ability of this organisation to branch and, within the proportions of the units, create partial and complete structural arches and loops which allow the system to span and fashion complex forms and spaces. Inspired by Roger Penrose's spin-network model, representing multi-linear states of interactions between particles and fields in quantum mechanics, S-String is a system that allows the construction of multiple load-path networks in space. This parallel illustrates how the study of networks as abstract mathematical-geometric organisations is the key to explaining structures, from that of the elementary particle to installations like S-String and larger structures.

1 AKT II, Canstruction installation, Canary Wharf, London, UK, 2012.
Plan sections through the installation show contact points between spiralling stacks of cans.

2 AKT II, Canstruction installation, Canary Wharf, London, UK, 2012.
3D model.

3

3 AKT II, S-String installation,
London, UK, 2011.
Truncated triangular prismatic
module.

4 AKT II, S-String installation,
London, UK, 2011.
Truncated triangular prismatic
module.

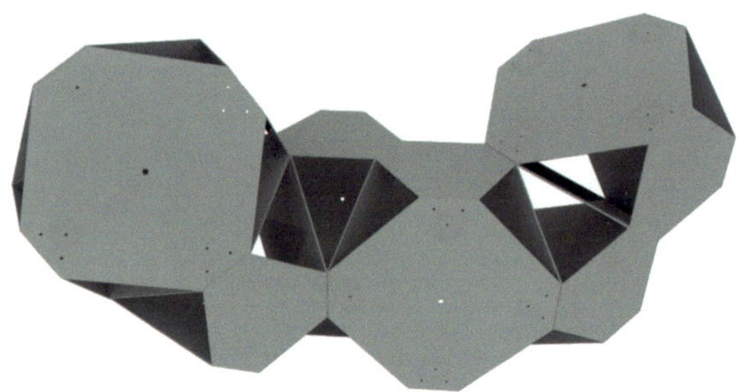

4

Organisational networks can be generated by the addition of components, as illustrated by the Canstruction and S-String installations, but they can also be obtained through subdivision. This is a recurring growth mechanism in nature, an example of which is the subdivision of a single fertilised seed or egg to more complex organisms. This recursive growth by subdivision creates emerging patterns in nature that can be described through simple mathematical rules and geometric proportions. One such example is the Fibonacci series, at the core of cellular division and characterised by golden proportions, closely linked to that same mathematical growth mechanism. Crystal Ceiling is an installation based on a three-dimensional extension of the Penrose tiling, theorised by mathematician Ludwig Danzer in 1991. Danzer's tiling (Figure 6) is based on four tetrahedral tiles creating a self-similar subdivision based on golden proportions. Further studies on the Danzer tiling showed that other aperiodic systems exist which allow the creation of layered planar surfaces. This development offers the opportunity to create double- and triple-layered structures composed of planes of polycarbonate sheets. The complex tiling never repeats due to its aperiodic nature, but it is composed of 16 tiles that subdivide into similar sets of the same 16 tiles, allowing modularity and repetition in the construction. In addition to this, every plane of polycarbonate is connected to another with identical clamps, due to the rigorous fivefold symmetry of the tiling system. Crystal Ceiling (Figures 7 and 8) is an example of a prototypical exploration in geometry, materiality and organisation that informs a wide range of larger projects, ranging from structural systems to facade concepts, and to materials and products.

Certain tiling patterns have particular properties of rigidity and strength. Reciprocal structures are formed by the assembly of discrete linear elements in interlocking arrangements, such that one element rests on the other until the perimeter is reached. Reciprocal structures have the benefit of achieving relatively large spans with short elements, connected by simple vertical bearing joints. Achieving continuity through discrete, short and transportable elements and with simple economic joints can be a benefit, as well as creating interesting 'interweaving patterns'. In the Palais des Banquets in Gabon (Figures 9 to 12), designed by Aranda\Lasch, the roof of the open-air theatre consists of a reciprocal system at different embedded scales. At the larger scale, the canopy is composed of seven different conical supports that fold horizontally to form the roof surface. These are interlocked in a reciprocal manner to create global lateral stability. Each conical support is composed of a reciprocal spiral of planar panels, forming a closed rigid system. Finally, each panel is assembled using a series of concentric reciprocal strips, creating rigid surfaces with the use of simple bolted connections. This pattern, composed of a constant-width strip, is produced in a single lightweight aluminium extrusion, cut to length and bolted together on site. This allows it to be packed and transported efficiently to the site and assembled with minimal works.

5 AKT II, S-String installation, London, UK, 2011.
Fully assembled installation at TestBed1, Battersea.

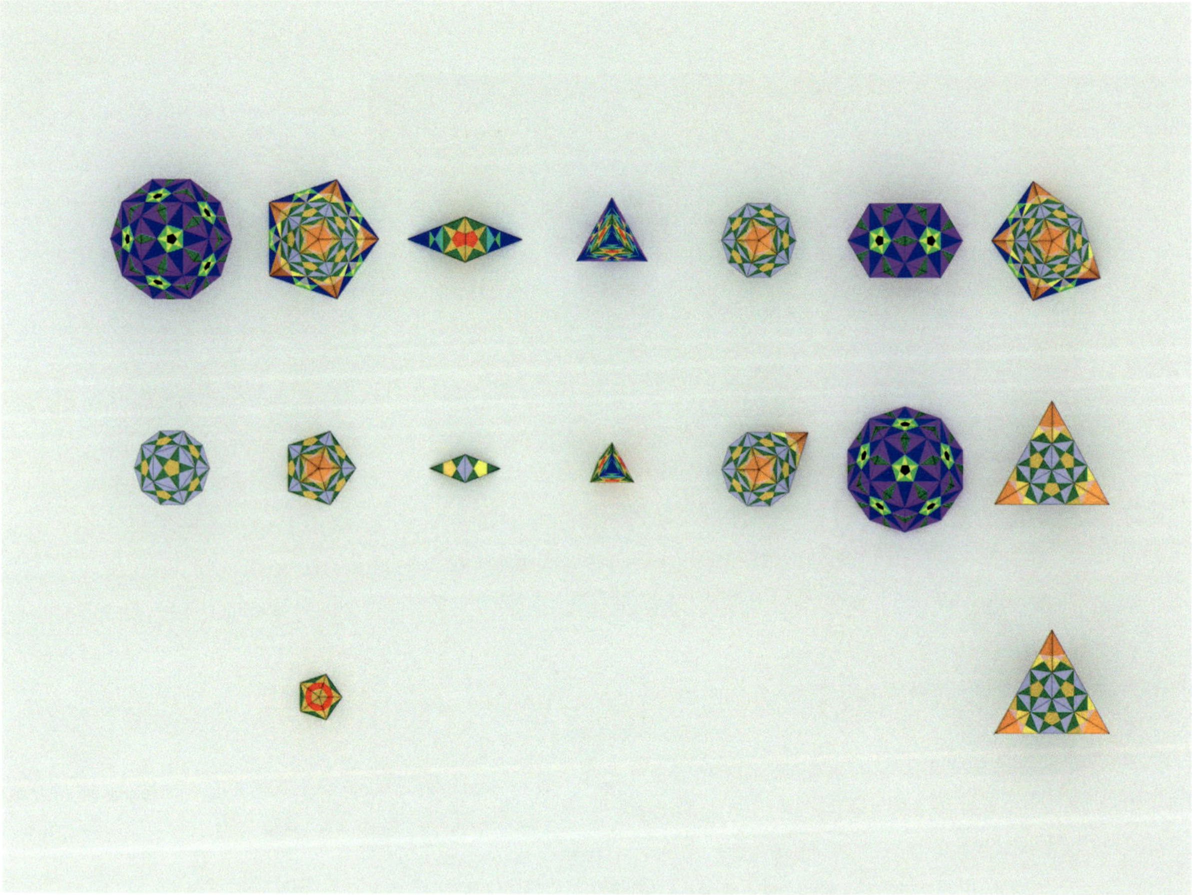

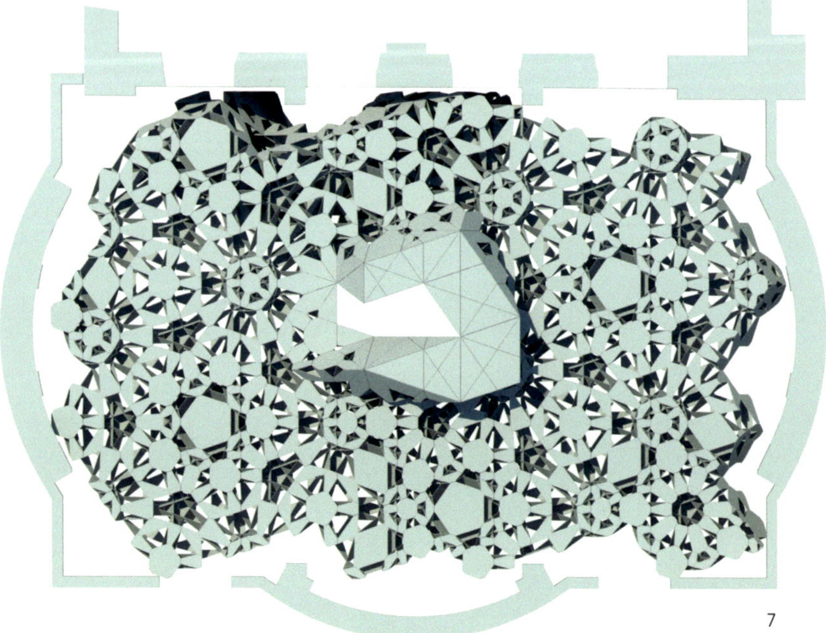

6 Balmond Studio with AKT II, Crystal Ceiling, Beauvallon, France, 2012.
The 16 tiles that constitute Crystal Ceiling and their self-similar subdivisions.

7 Balmond Studio with AKT II, Crystal Ceiling, Beauvallon, France, 2012.
Plan of Crystal Ceiling, consisting of a 3D aperiodic tiling pattern.

8 Balmond Studio with AKT II, Crystal Ceiling,
Beauvallon, France, 2012.
Prototype of a section of Crystal Ceiling,
composed of polycarbonate sheets and
standard polycarbonate connections.

The Coca-Cola Beatbox pavilion, in the London 2012 Olympic
Park (Figures 13 to 15), is also based on the reciprocal interlocking
of rigid elements, forming a self-supporting structural facade of
5 x 1 m rectangular ETFE cushions. However, unlike a traditional
reciprocal system, this is composed of 3D interlocking elements in
space, providing a vertical and horizontal load-bearing structure.
The reciprocal tiling of the ETFE cushions – which are reinforced
by a steel rectangular frame – creates a regular network of tangent
connection points, allowing cushions and connections to be
geometrically identical and therefore economic to build and easy
to assemble. While based on a regular grid of connection points,
cushions are allowed to slide and flip 90° about their longitudinal
axis to create an irregular arrangement. This installation was far
more than a structure; it was an experiential, interactive, responsive
sound and light system. The play of its geometry and pattern
created its intended seemingly chaotic rhythm with a rigorous
underlying organisation that allowed its quick and economic
realisation on site.

From the piling of brick-like cans to the divergence of particular
polyhedral modules into branching structures, from the creation
of continuous space frames of aperiodic tiles to the construction
of horizontal and vertical reciprocal interlocking of elements, the
above examples demonstrate how structure is created through
the organisation of its components, the topology and pattern of its
connective network and its underlying geometric system. However,
these principles also extend to continuous structures. Even when
cast in smooth monolithic concrete surfaces, these are poured in

9 Aranda\Lasch with AKT II, Palais des Banquets roof canopy, Libreville, Gabon, 2013.
The reciprocal interlocking of the roof structure gives stability to the structure.

10 Aranda\Lasch with AKT II, Palais des Banquets roof canopy, Libreville, Gabon, 2013.
The reflected ceiling plan incorporates lighting strips.

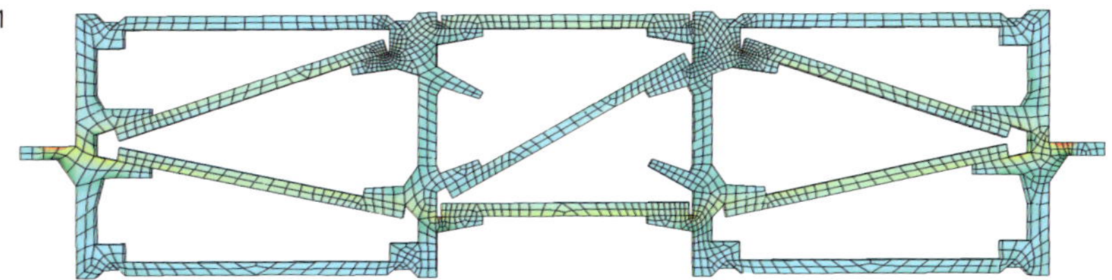

11 Aranda\Lasch with AKT II,
Palais des Banquets roof canopy, Libreville,
Gabon, 2013.
Finite element model of an extruded aluminium
panel. This is composed of bolted extruded
aluminium profile into 1.5 m wide panels.

12 Aranda\Lasch with AKT II, Palais des
Banquets roof canopy, Libreville, Gabon, 2013.
View of open-air theatre stage with background
of Gabonese landscape.

shutters of discrete ply sheets and therefore have engrained within their surfaces the patterns of their formwork, and their armature is composed of discrete mats of reinforcement. When realised in steel monocoque (Figures 16 to 18) or timber stress-skin construction, continuous structures are characterised by weld lines and stiffener patterns and, when using orthotropic materials such timber and carbon fibre, the direction of the fibres within the material dictates the principal load-paths.

In the past decade, we have seen structures like the Arnhem Centraal station, the Taichung Opera House and the Bournemouth Art Studio being completed on site and, more recently, the Algiers Presidential Palace and the Al Fayah Park project being designed and developed. These are all based on the use of NURBS surfaces, the geometric structure of which, in Euler's terms, was used as the basis for the physical, material, structural organisation and articulation. Based on the use of the Catmull-Clark subdivision algorithm, which allows the computation of smooth fine quadratic meshes starting from their crudest topological equivalent, bespoke tools were developed to create continuous NURBS polysurfaces. These smooth polysurfaces of simple quadratic NURBS patches are characterised by continuous UV lines running through them.

Reciprocal tiling

Proliferation of the tiling pattern

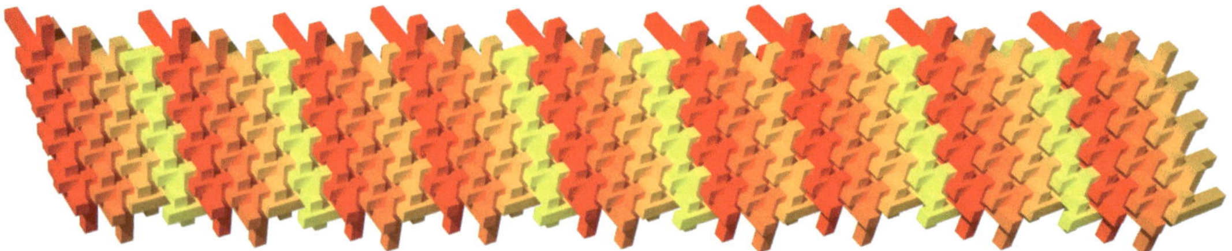

Rhythm of the tiling pattern

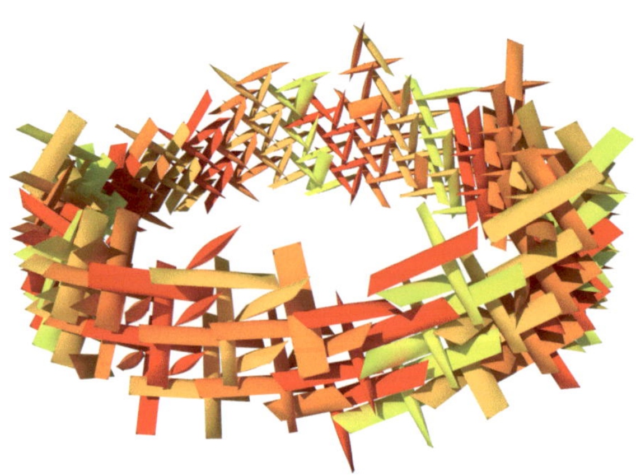

Mapping of the pattern on cylindrical form

13 Pernilla & Asif with AKT II, Coca-Cola
Beatbox, Olympic Park, London, UK, 2012.
The reciprocal interlocking module is composed
of three tangent ETFE pillows.

14 Pernilla & Asif with AKT II, Coca-Cola
Beatbox, Olympic Park, London, UK, 2012.
Standardised connections between ETFE
facade components.

15 Pernilla & Asif with AKT II, Coca-Cola
Beatbox, Olympic Park, London, UK, 2012.
The assembly of ETFE facade components was
undertaken on site.

14

15

A first-principles approach to geometry and topology, where the organisation and inner logic of a structure is understood and exploited, opens new opportunities to design engineers and their collaborators in the creation of complex contemporary buildings that have to respond to increasingly complex human needs, while being as economic and sustainable as possible. Linear Cartesian systems have characterised buildings through classical and Modernist architecture; frames of beams and columns have been the norm since the standardisation of building structures from the advent of the Industrial Revolution. These often still offer effective solutions, but should not be the limit to the imagination, creativity and ingenuity of designers of this and future generations.

As the definition of geometry in contemporary mathematics has gained a more generic and abstract meaning, we as design engineers today use the understanding of complex geometric systems to inform the design of complex dynamic structures and the environment around them, such as CFD[1] analysis at Centre Point in London (Figure 19). This has blurred the boundary between the structure and organisation of a system and its geometric configuration in space. In representational terms, the algorithm that describes a certain organisation often becomes a synonym for the form it produces. Our bespoke tools today

16

17

18

16 SimpsonHaugh and
Partners with AKT II,
Library Walk Link Building,
Manchester, UK, 2015.
Fabrication of 6 mm
monocoque stiffening cage
by CIG.

17 SimpsonHaugh and
Partners with AKT II,
Library Walk Link Building,
Manchester, UK, 2015.
Welding of 12 mm stainless
steel structural skin by CIG.

18 SimpsonHaugh and
Partners with AKT II,
Library Walk Link Building,
Manchester, UK, 2015.
Assembly of monocoque
structure on site.

19 AKT II, Centre Point,
London, UK, 2014.
CFD analysis of dynamic wind
effects.

embed complex analytical processes within synthetic intuitive forms, allowing the process of mathematical rigour to merge with that of intuitive creation.

REFERENCES
1 Computational fluid dynamics, usually abbreviated as CFD, is a branch of fluid mechanics that uses numerical analysis and algorithms to solve and analyse problems that involve fluid flows. Computers are used to perform the calculations required to simulate the interaction of liquids and gases with surfaces defined by boundary conditions. With high-speed supercomputers, better solutions can be achieved. Ongoing research yields software that improves the accuracy and speed of complex simulation scenarios such as transonic or turbulent flows. Initial experimental validation of such software is performed using a wind tunnel with the final validation coming in full-scale testing, eg, flight tests. (Wikipedia, https://en.wikipedia.org/wiki/Computational_fluid_dynamics)

TEXT
© 2016 John Wiley & Sons Ltd

IMAGES
Figures 1, 2, 3, 4, 5, 6, 7, 9, 11, 13 and 19 © AKT II; figure 8 © Balmond Studio; figures 10 and 12 © Aranda\Lasch; figures 14 and 15 © Ed Moseley/AKT II; figures 16, 17, 18 © CIG Architecture

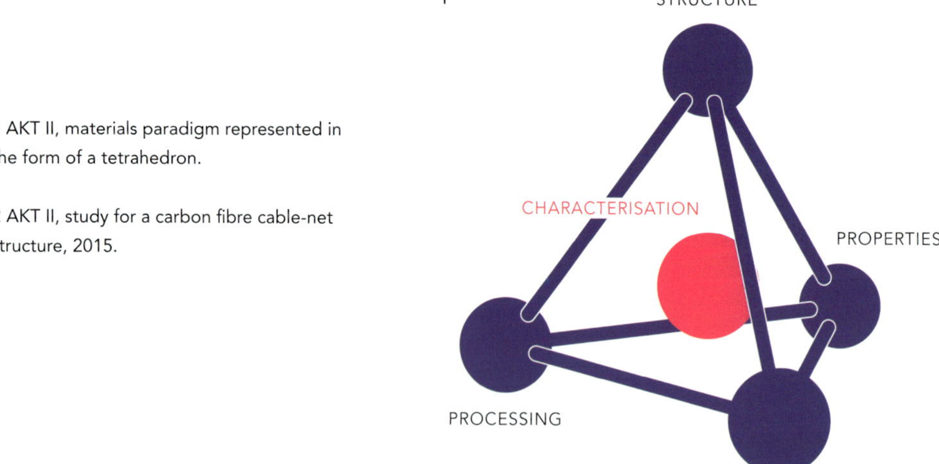

1 AKT II, materials paradigm represented in the form of a tetrahedron.

2 AKT II, study for a carbon fibre cable-net structure, 2015.

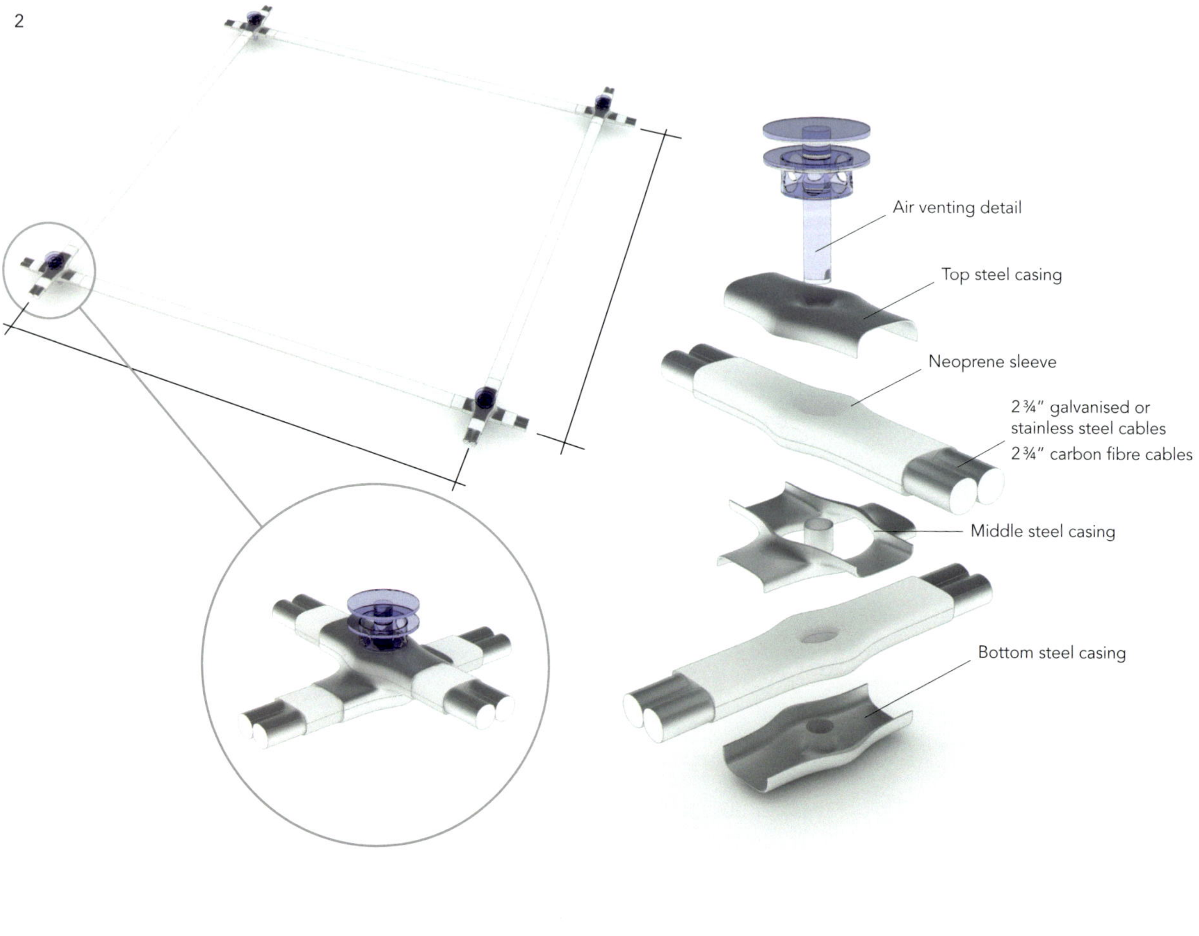

9 MATERIAL MATTER

ED MOSELEY AND
MARTIJN VELTKAMP

In recent years, advances in the research of new materials, their properties and applications, have placed an emphasis on material science as an emerging and interdisciplinary field that draws from chemistry, physics and biology, and feeds into engineering applications. Breakthroughs in material science have led to an escape from the limitations of available materials, and are having a significant impact on the development of new technologies. Engineers have had to sharpen their lens, peer deep into the composition of materials and forensically analyse their physical properties. With the research and certification of new materials and the application of new technologies, the palette of materials available to construction is expanding.

While fabrication techniques and the performance of building components are explored at the building scale, the processing and performance of a material are similarly tested in the lab (Figure 1). Through the corroboration of computer models with physical tests, computer simulations of building materials are becoming more reliable, at both the microscopic and macroscopic level.

Today, areas of material innovation in the construction industry include new methods of forming and manufacturing traditional structural materials such as steel and concrete, as well as composite and synthetic materials with enhanced properties. Certain materials, such as carbon fibre (Figure 2) or Kevlar, display higher strength-to-weight characteristics, making them structurally efficient, while others score highly on multiple performance parameters, transmitting daylight or insulating from heat. Use of these new materials is always subject to their cost, a parameter in large applications like buildings.

Sometimes, the best ideas come from maximising what is already available. An example of this is fabric-formed concrete, where wet concrete is no longer cast into rigid moulds, but is instead cast against fabric, resulting in organic forms. This can be many times cheaper than conventional formwork. While this method is currently in its infancy, the potential to deliver greater material efficiency, as well as a relatively low-tech construction method, is already appealing to architects. In addition to this, it may solve one of the other great challenges of concrete: durability. The permeable formwork of fabric-formed concrete allows free water to soak away, leaving a cement-rich, durable outer shell on the concrete.

3

4

3 SimpsonHaugh and Partners with AKT II, Library
Walk Link Building, Manchester, UK, 2015.
Form-finding of monocoque soffit geometry.

4 SimpsonHaugh and Partners with AKT II, Library
Walk Link Building, Manchester, UK, 2015.
Assembly of monocoque structure on site.

5 SimpsonHaugh and Partners with AKT II, Library
Walk Link Building, Manchester, UK, 2015.
Structural glass facade and monocoque roof structure
diagram.

In monocoque and stress-skin construction, traditional materials are used to serve different purposes, from the definition of the external form and appearance, to the assessment of its structural strength and stability, to the enclosure of internal space from the environment. In some cases, like the Manchester Library Walk Cloud (Figures 3 and 5), this has reduced the number of different layers and components required to construct a building, avoiding the need for separate structure and building envelopes that require complex and sometimes unnecessary coordination and investment of material resources. In the Cloud, the polished stainless-steel monocoque roof is the reflective architectural form – the structure and the waterproof enclosure being composed of prefabricated panels welded to form a seamless object. Similarly, the vertical glazed wall is a triple-layer, frameless structural glass support to the roof as well as the enclosure.

Efficiency in the use of material is certainly key to the economy of a structure, extending beyond financial gain. Reducing the investment of material resources and the energy required for their manufacture has a critical impact on the sustainability of structures, given the large proportion that the construction industry contributes to the carbon footprint. In addition, more efficient buildings use biologically renewable materials such as timber, which helps to reduce the impact on the environment. An example of this is the use of up-cycled timber in products such as cross-laminated timber (CLT) and its use within lightweight building components, for example prefabricated floor slabs (Figures 6 and 9). This offers several benefits, from intrinsic fire resistance to acoustic insulation properties.

MATERIAL WORLD

Assessment of the structural properties of materials and system forms is an essential part of any structural engineer's work. During a project, an engineer will utilise codes and tabulated properties to develop a system for a primary structure or building envelope. For many projects this is appropriate, as deviation from the conventional carries an inherent risk, but when projects seek potential benefits from a non-standard approach, the level of risk is accepted into the design process; the attitude and approach of the team (client, architect, engineers and contractors) becomes an essential part of making the project work. Part of an engineer's role is to assess and manage that risk, but if this risk is apparent, uncertainty permeates through the team. This can be common in a large-scale project, where a plethora of conventional solutions is available, and where a committee is responsible for the project's success. In contrast, an individual or small team's decision to take a risk can be made more unilaterally.

There are, however, benefits to using physical testing to justify a design. With more conventional materials, a different use or system approach can be justified, allowing a non-typical use to be adopted without the need for excessive design. When less conventional materials are proposed (materials not typically considered as structural), it is usually to enable some form

5

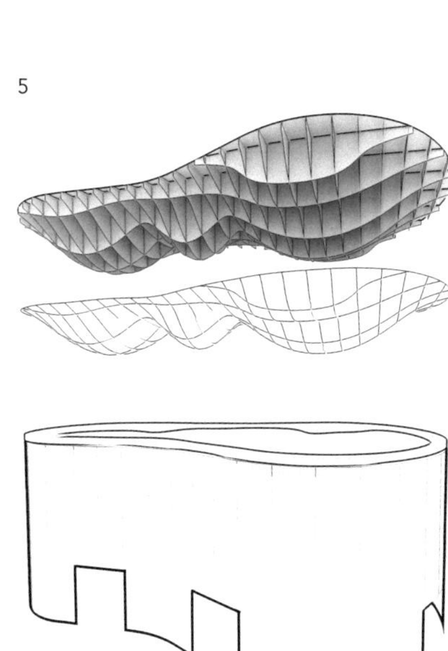

6 AKT II, 2015.
Study for a prefabricated modular floor system
using a composite CLT/steel-tray system with an
accessible raised timber floor.

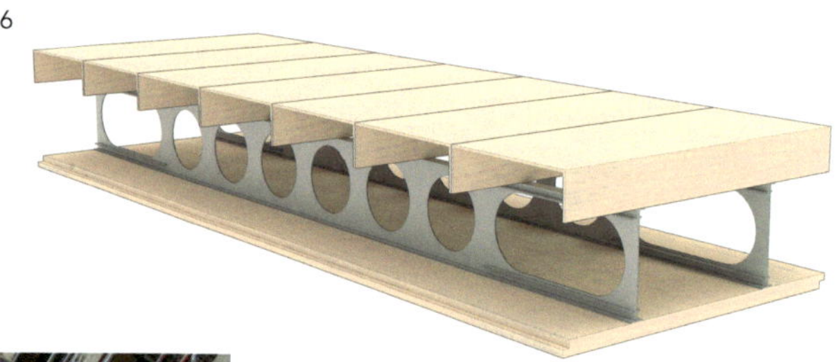

7 Stanton Williams Architects with AKT II, Aga
Khan student residence block, King's Cross,
London, UK, 2015.
View of the hit-and-miss brickwork which was
verified by the physical testing process.

8 Stanton Williams Architects with AKT II, Aga
Khan student residence block, King's Cross,
London, UK, 2015.
Internal view of hit-and-miss brickwork, no wind
posts were required in the exposed areas, as
verified by physical testing.

9

of polyfunctionality, to use the properties of the material to provide a non-structural aspect of the design proposal, or in response to an alternative design parameter. In such a case the structural performance is a secondary concern. While mock-ups of architectural features are commonplace in many projects – allowing a client to make decisions on facades, finishes, etc – the idea of testing structural systems for the betterment of the project doesn't feature as frequently. Perhaps there is an expectation that everything about how a building's structure will behave is already known, or perhaps aesthetic assessments are more tangible to non-technical members of a team. However, systems can behave differently from predictions based purely on material properties.

The testing of conventional structural materials for specific roles has been a feature of the work done for pavilions and installations over the years, and has also been used to verify various systems for large-scale projects. A hit-and-miss brickwork facade system was designed for the Aga Khan student residences at King's Cross (Figures 7 and 8) – a project delivered with Stanton Williams as architect – which builds on the work of such innovators as Eladio Dieste, modern academics such as Philippe Block, Fabio Gramazio and Matthias Kohler, and more commercial pioneers.

The proposal looked to use the brick facade to create a window with the multi-functional role of controlling light levels and visual privacy, while also developing the architectural aesthetic for the proposal, all with the aim of producing a building that feels like a traditional, natural structure. Four options were considered to achieve the desired brick arrangement. The testing process was performed for two of these options; one of which was used in the final construction. Inflatable air bags were used to pressure-test a sample brick panel to verify that it could resist the potential wind loadings. Due to the significant amount of 'miss' in the design, windposts are usually a codified 'must', but this would have had a detrimental effect on the views from the bedroom windows. During testing, a pressure of 4.0 kPa was applied to the hit-and-miss system. While this seems very high relative to the predictable potential loadings, testing of this nature needs to account for material variation and the potential for inconsistent workmanship during construction. The testing confirmed that the most economical and rapid construction solution could be used, and verified that a shear failure in the bed joints could potentially occur at less than 0.4 kPa.

The term 'structural materials' has become synonymous with masonry, steel, concrete, aluminium, glass and timber, while other materials are rarely considered; occasional research into alternative structural materials seldom gets beyond being used for a single installation, only rarely making it to the status of a 'pavilion'. In terms of scale of construction using alternative structural materials, little has progressed beyond what was achieved with the Monsanto House of the Future in 1954 (Figure 10). The traditional exception to the normal design routes (exhibitions, pavilions and follies) enables materials' function to be defined by their properties, as

10 Marvin Goody & Richard
Hamilton, Monsanto House
of the Future, Disneyland,
California, USA, 1957.

10

11

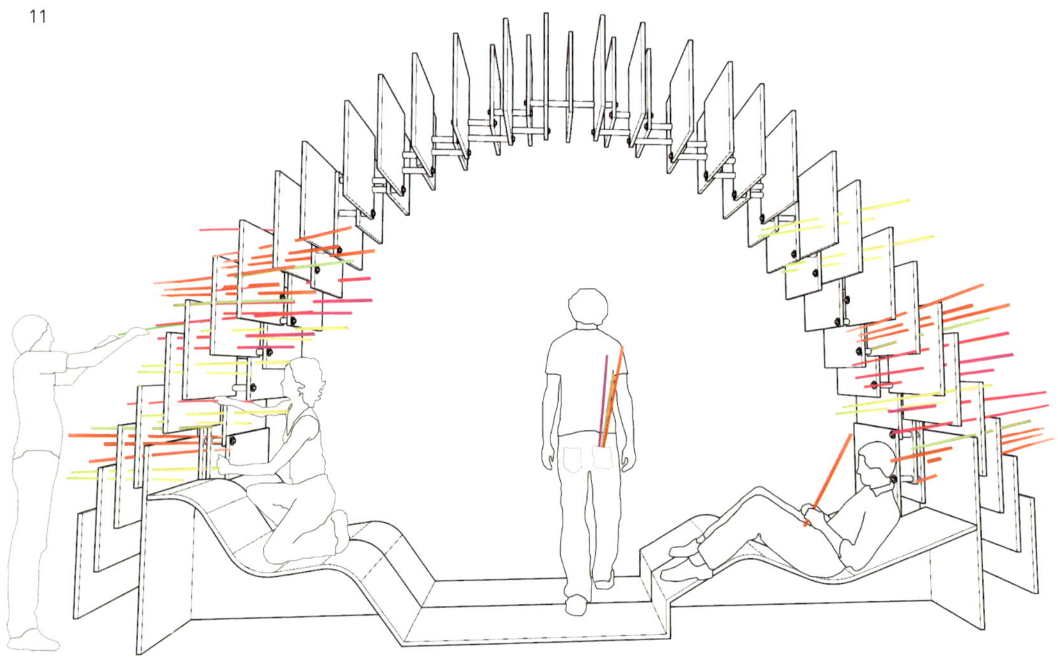

the control is within a smaller group, and the ambition allows for the unconventional if it benefits the project: scale is a big part of the issue. During the design development of most commercial projects, a schedule of responsibilities and a design responsibility matrix are developed to split the grey area of building behaviour into black and white. Some of the materials are defined as primary structure and sit with the structural engineer; others, such as facade and design responsibility, sit with the facade consultants and suppliers, and yet others are frequently the responsibility of the architect and designated as finishes or partitions. This division of responsibility is frequently done without cognisance of the proposed material, its mechanical performance properties, or how it could be utilised within the building behaviour to maximise efficiency in material use.

The approach of bespoke testing materials to enable unconventional structural components to be used was taken forward during the Royal Academy's 2014 exhibition, *Sensing Spaces* (Figures 11 to 15). The proposal by Berlin-based architect, Diébédo Francis Kéré, was to utilise polycarbonate panels to form a tunnel with a varying arch profile. Its polycarbonate properties as a material are known, but here the honeycomb panels were constructed by heat pressing the polycarbonate tubes together. This bond defined the panel's flexural performance, particularly at the connections, so some empirical testing was needed to get sufficient data to use this in an analysis model, and also to assess a potential failure stress for the bond. The original concept utilised multiple stiff rods to reduce bending stresses, but it was agreed that flexible rods were required to enable construction and account for misalignment between holes in the panels. After a full-scale mock-up was built with nylon rods, the structure was proved to be unstable, even after retesting with fewer rods.

Instead, a new approach which required denser packing was developed to reduce the effects of flexure and enable the compressive action of the arch to be the dominant influence. Due to variations in the arches' structural arrangements, there were still flexural movements and significant predicted movements; however, the effect of multiple arches gave robustness to the final form not found in the individual arches. Working with materials that are not typically seen as structural necessitates a consideration of the properties that lead them to be considered unusual: in this case, high levels of flexibility and sudden loss of capacity at the connections when the heat-pressed bond failed.

Similar packing considerations were used in the development of the [C]space Pavilion, built with the DRL at the Architectural Association in 2008. This used an interlocking system to create the fundamental principles of the geometric arrangement, allowing the basic structural principles to be defined as part of the model generation and then the analysis to define whether this would be a two-layer or three-layer piece of fibre-reinforced plastic. While the geometric practice and the analysis and testing process for this pavilion were similar to those used successfully in the *Sensing*

11 Kéré Architecture with AKT II, Sensing Spaces exhibition, Royal Academy of Arts, London, UK, 2014.
The original concept for the arch was to create an interactive experience that could grow through the visitor input during the period of the exhibition.

12 Kéré Architecture with AKT II, Sensing Spaces exhibition, Royal Academy of Arts, London, UK, 2014.
Reciprocal interlocking units.

12

13

13 Kéré Architecture with AKT II, Sensing
Spaces exhibition, Royal Academy of Arts,
London, UK, 2014.
Image of the Rhino model and reciprocal
interlocking units that were used to create
the installation.

14 Kéré Architecture with AKT II, Sensing
Spaces exhibition, Royal Academy of Arts,
London, UK, 2014.
Testing of the proposed honeycomb panel
to be used in the formation of the arch
structure.

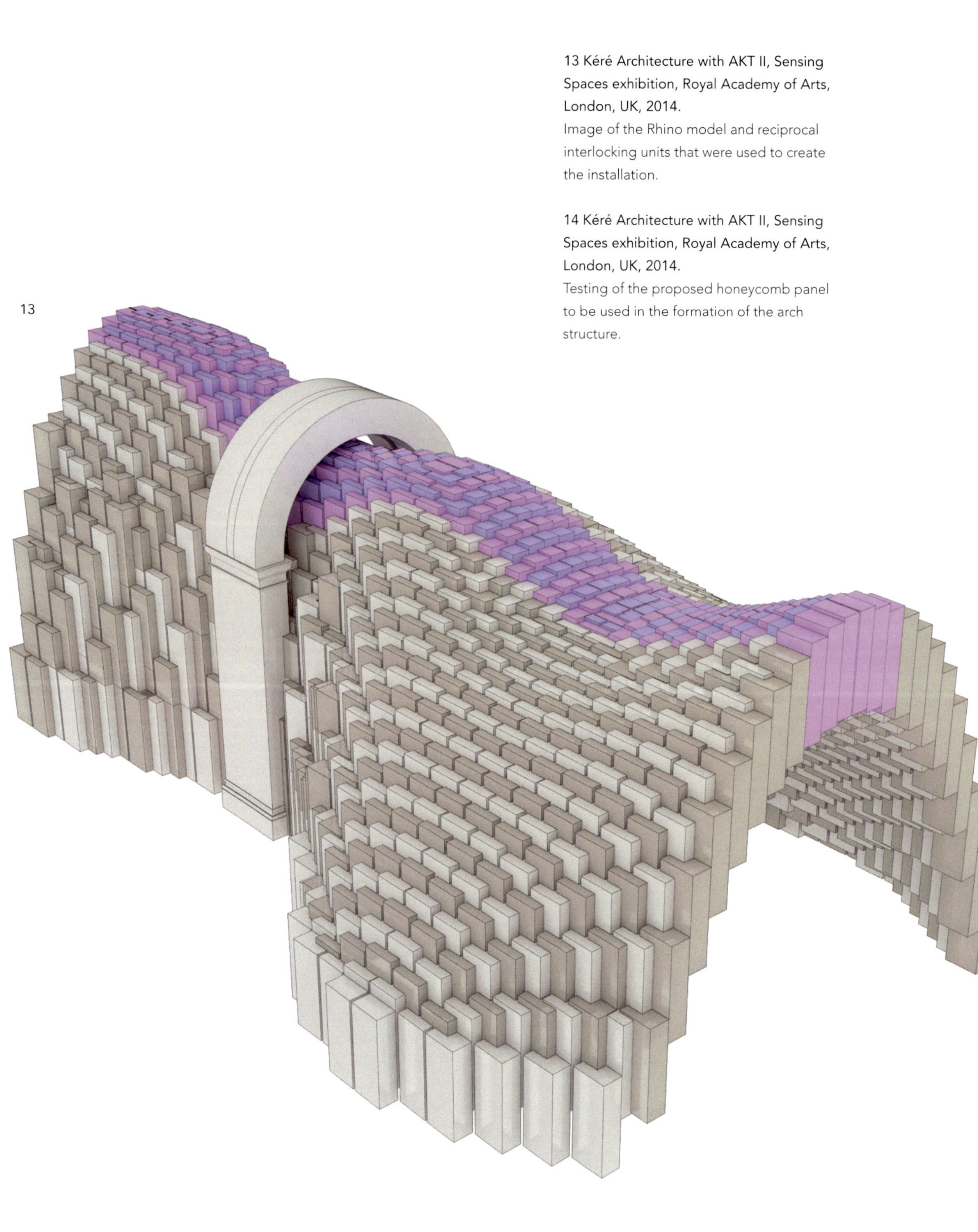

14

Spaces exhibition, the pavilion benefited from being constructed from a tried-and-tested structural material: concrete. This assisted with the approval process that enabled it to be a public pavilion. When developing the UK Pavilion for Shanghai Expo 2010, however, a much more conventional, column system was proposed, but this time the materiality was more unusual – the proposed legs were to be acrylic. Despite the testing put in place to verify that the acrylic legs could support the pavilion proposal, the final construction used steel legs, primarily due to a lack of comfort in using unconventional materials as structural elements.

Sensing Spaces benefited both from being on a smaller scale, and by being branded an exhibition as opposed to a pavilion, which produces different expectations. If the installation were bigger and perhaps located outside a building, it is questionable as to whether the same level of design and testing would have provided sufficient confidence for the project to be realised.

FIBRE-REINFORCED POLYMERS

Fibre-reinforced polymers (FRPs) are a hybrid, man-made, engineered material of structural fibres encapsulated in stiff polymers, which offer many advantages in a wide variety of design requirements. They belong to a family of tailor-made materials with a high stiffness-to-weight ratio, stable characteristics in time and the ability to be moulded into complex shapes. Given time, their properties are stable, without the need for maintenance. The structural integrity of the material is due to the type, orientation and number of fibres inside it, as well as the type of matrix material. Glass fibres in a thermoset polyester matrix are cost-effective, while carbon fibres set in an epoxy matrix offer more strength and stiffness, but at a higher cost.

Various processes exist to mix the fibres and resin, each with their own specific advantages in relation to pronounced shapes, standardised elements and size limits. In the resulting products, whatever the manufacturing process, it is the fibres that create structural integrity. The ratio of fibres to resin determines the eventual mechanical properties, as well as the weight of the polymer. As much as with any other structural material, understanding the relationship between the end product and material production is instrumental for good design.

FRPs were first applied in the Second World War, and subsequently used in defence and aerospace for their strength in combination with their lightweight and radar-transparent properties. These properties could be adjusted for each specific situation (whether for strength, stiffness or resistance to extreme conditions), while always keeping the material's ability to be made in any form. FRPs were also applied in car bodies, where low weight-per-unit stiffness and weight-per-unit strength are equally important (Figure 16).

In buildings, FRPs were used in post-war architecture to construct facade elements or entire buildings such as Matti Suuronen's Futuro Houses, which served as relocatable ski huts. The material

15

16

17

attracted new interest with the free-form design trend in the early 21st century, when the rise of digital modelling technologies enabled designers to define such shapes, putting pressure on the construction industry to physically realise them. The rise of numerically controlled 3D milling for the mould, in combination with FRPs to provide strength and stiffness, made these applications possible.

Due to the economic downturn and reduced construction budgets, the free-form era is past its prime, but FRPs are still very much in use today and notably appreciated for their longevity. Insusceptibility to corrosion, rot or UV radiation, and subsequent low-maintenance requirements, make FRPs a particularly interesting material. So they have also found their way into civil-engineering applications, such as bridges and lock gates, both requiring long design lives where maintenance is cumbersome and costly (Figure 17). FRPs, therefore, bring qualities that corrosion-prone steel and concrete – which suffer from penetrating chlorides – can't beat. In refurbishment projects, an FRP bridge deck offers higher load-bearing capacity than the steel deck it is replacing, but at a much lower weight. This means that elements can be prefabricated in large sizes, since weight is not the logistically critical parameter any more – the dimensions are.

Aerospace, defence and Formula 1 have all brought about prototype applications and technological advancements. After lengthy research, the initial high unit costs are eventually pushed down by serial production as they are brought into everyday applications. For example, all rotor blades in wind turbines are realised in FRPs, and tidal energy turbines will soon follow.

In aerospace, FRPs are now improving aerodynamics, since doubly curved shapes can now be built as structural forms, reducing both fuel consumption and CO_2 emissions. These aerodynamic forms will soon be applied to long-span bridges to minimise wind loading, and mitigate any adverse dynamic interaction effects between wind and structure. The same aerodynamic advantages achievable through the use of FRPs apply to high-rise buildings, where aeroelastic effects influence the performance of structure, and airflow at street level affects pedestrian comfort.

When compared with reinforced concrete – another man-made material that relies on strands of reinforcement in different directions within a binding material – FRPs can be manufactured in much thinner sheets, due to the stronger, finer fibres and the ability of the resin to bond without the need for deep cover to the reinforcements. FRPs allow robust and orthotropic properties to be exploited in complex and varying loading conditions.

Multiple load-bearing directions are also present in laminated veneer lumber (LVL), but whereas LVL is a standard product available in specific thicknesses and dimensions, FRPs are always bespoke and can create durable and fully water-resistant synthetic equivalents of timbers, curved and folded, comprising

18 Synthesis Design + Architecture, dismountable pavilion for the introduction of the Volvo V60, Milan, Italy, 2013.
The pavilion is all-structure: an FRP tubular edge member was produced straight and then bent to follow the pavilion's tensile web.

19 FRPs are used extensively in the Airbus A350 XWB, and during its first flight it was provided with a carbon fabric-style livery to illustrate this.

20 Institute for Computational Design (ICD) and Institute of Building Structures and Structural Design (ITKE), ICD/ITKE Research Pavilion 2012, University of Stuttgart, 2012.
ICD-ITKE's pavilion was built with a few planar rigs as its only mould.

various grades of fibres and joined in multiple directions with no intermediate joints.

FRPs suit all geometries, including those that are non-orthogonal, since the fibrous raw material does not have a fixed form and can be laid out into any shape. These are made rigid through the infusion of resin into the fibres, which, once cured, gives the material strength and stiffness. While FRPs do require a mould, they don't need to withstand high pressures or a significant self-weight, so the moulds can be simpler and more cost-effective: mouldings can be as simple as a foam block like those used in electronics packaging. Alternatively, they can be vacuum formed upside down into the mould.

The fibres come in standard sheets that are laid in the mould by simply unrolling them. To ensure straight fibres and avoid wrinkles, the geometrical rules of developability are applied at the material level of scale. Fibres in multiple directions can be combined within a fabric, offering a degree of prefabrication. The fabric's stitching may allow rotation of fibre directions, thus enabling double curvatures.

In aerospace construction, precise computer-controlled placement of strands of fibres is common. ITKE-ICD's pavilion is an experiment in eliminating the need for a mould, using simple planar rigs spun around as fibres are laid by robotic arms (Figure 20).

While FRPs offer extreme versatility in manufacturing, their design, specification and procurement are highly specialised. Geometry, pattern, structural performance and fabrication need to be considered simultaneously from the earliest stages of design. For this material to be considered in widespread building applications, it requires designers and builders to understand each other's working processes. For example, standardisation can reduce complexity and cost, but may limit versatility in the design. FRPs are not a new material, but the acceptance of their widespread use in building applications requires a more sophisticated and integrated process of design, engineering and production.

TEXT
© 2016 John Wiley & Sons Ltd

IMAGES
Figures 1, 2, 3, 4, 6, 9, 12, 13, 14 © AKT II; figure 5 © Valerie Bennet/AKT II; figures 7 and 8 © Stanton Williams 2016; figure 10 © Ralph Crane/Getty Images; figure 11 © Kéré Architecture; figure 15 Photo © Royal Academy of Arts, London, 2014. Photography: James Harris./© Kéré Architecture; figure 16 © General Motors; figure 17 © A. Gortemaker; figure 18 © Synthesis Design + Architecture; figure 19 © Airbus, photo Sylvain Ramadier, 2014; figure 20 © ICD/ITKE Stuttgart University

10 STRUCTURAL DYNAMICS

PHILIP ISAAC

The field of structural dynamics has undergone an almost unrecognisable transformation over the last forty years, driven by the exponential increase in computing power seen over this time, the continual drive for greater efficiency in structures, and the greater understanding of previously unknown dynamic phenomena. For modern design engineers, dynamics can no longer be treated with trepidation and dealt with through the broad brush of conservatism. Instead, they must harness the tools at their disposal to be proactive, rather than reactive, during the design process. They must also be able both to grasp the complex analysis dynamic situations often require, and to communicate the results and implications effectively to the rest of the design team.

This chapter starts by investigating the historical and mathematical context of dynamics, building up a picture as to why dynamics is becoming more and more important in design before introducing specific projects where unique solutions were required to solve dynamic issues. The examples used in this chapter mainly focus on vibrational analysis, which is the most often encountered dynamic situation in structural engineering. However, the same theory that underpins this can also be applied to other aspects of structural engineering, and the final part of the chapter will take a brief look at these areas.

As structures continue to push the boundaries of what is physically possible, a major part of the work design engineers carry out is dynamic analysis. As materials are used more efficiently, lengths of spans increase, and structures get lighter, dynamic issues become more and more prevalent. Dynamics, though, is not a new phenomenon. Indeed, it could be considered that all loads applied to a structure are in some way dynamic; however, to an engineer, whether or not a load should be considered dynamic is a question of relativity. In situations where the frequency of the load is similar to the natural frequency of vibration of a structure, dynamic effects are likely to be relevant. This covers many situations: the action of wind on buildings and bridges, the pressure from an explosion, human-induced vibrations, the action of earthquakes and vehicle loads on bridges.

The theory which underpins modern structural dynamics is not new; the mathematical theory to determine the natural frequency of a structure was first demonstrated by Jean d'Alembert in the

1 AKT II, classic dynamic model.
A simply supported beam is converted to an equivalent mass/spring system.

2 AKT II, the principles of divergence in the context of bridge design.
Wind-induced dynamic torsion of the bridge increases the loaded area, exacerbating the dynamic forces on the bridge.

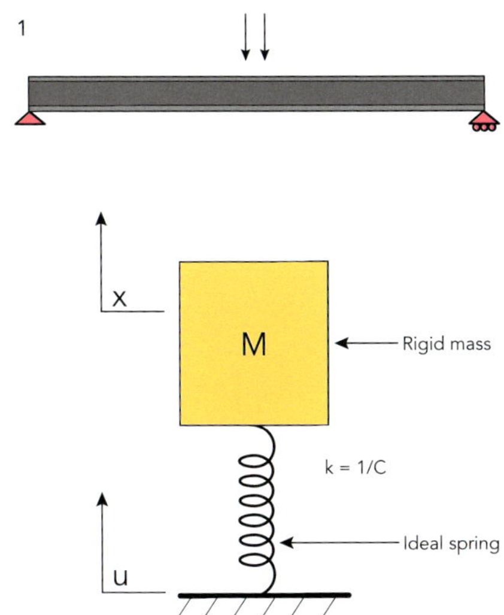

1

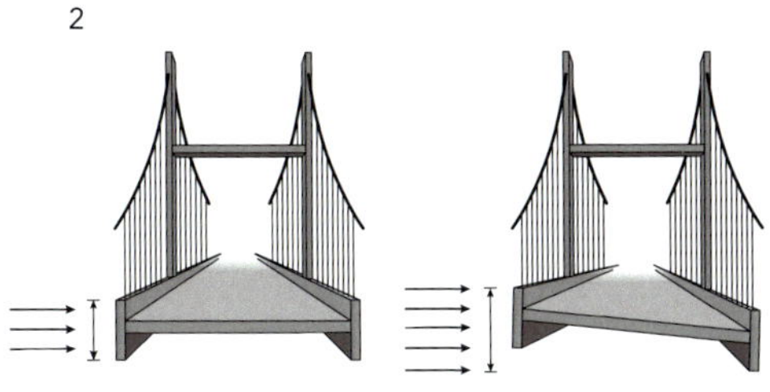

2

18th century. It showed that for a single mass/spring system (Figure 1), the natural frequency was related only to the stiffness and mass of a structure. While the solution for a one-degree-of-freedom (DOF) structure is relatively simple, the complexity is due to modern structures having not one, but many thousands of DOFs. As each DOF has its own natural frequency of vibration, it is clear that solving all of the equations to find all of the natural frequencies could become, even for a linear elastic structure, very difficult indeed.

Today the power of modern computers, coupled with the advancement of commercial finite element analysis packages, has meant that determining the natural frequencies for even the most complicated structures can be achieved relatively quickly on a desktop computer. This is in great contrast to the situation thirty or forty years ago when such a computer analysis would have taken many months to complete and would have usually cost a prohibitive sum of money. It can therefore be considered that improvements to analysis techniques have allowed for much greater material efficiency in design by giving engineers the confidence that lighter, longer and more daring structures – which would previously be considered dynamically impossible – are in fact possible.

There are, however, instances where these limits have been pushed too far and previously unknown or unforeseen dynamic phenomena have occurred (Figure 2), leading to collapse of a structure or significant issues with dynamics. Two notable examples of this are the Tacoma Narrows Bridge (Figure 3) and the Millennium Bridge for pedestrians in London (Figure 4). The former quickly picked up the nickname 'Galloping Gertie', as from the beginning it suffered from an aeroelastic effect known as 'divergence'. This occurs when the area of a structure which catches the wind actually increases as it moves, causing the force acting on the structure to increase. At the time, the phenomenon causing the failure was not well-understood and, as a result, many bridges were retrofitted with deep trusses to stiffen the structures. Over time, however, as aeroelastic phenomena have become better understood, bridge design has become more efficient and issues with wind can now be dealt with in a more scientific and efficient manner. In the case of the Millennium Bridge, a little-known phenomenon referred to as 'lock-in' almost caused failure. In this case, the bridge underwent noticeable deflections as people walked across it. As people stepped laterally to counteract the bridge's motion, they ended up walking in sync with one another. This had the effect of amplifying the horizontal forces on the structure, causing it to 'wobble'.

It is vital for the design engineer to learn from these important lessons, and to use the experience gained to respond quickly and reliably when new challenges present themselves in design. With the abundance of tools at their disposal and the wealth of experience gained over the last half-century, the challenge for modern design engineers is how to harness these tools and use them effectively within the design process.

3

An example of implementing these lessons is the design of a six-storey, 180 m long continuous ramped staircase in Bloomberg's new headquarters in central London. The plan layout was for a three-pointed hypotrochoid curve (Figure 6), with each flight between landings following a sweep of 30 m with an intermediate support at the mid-span. From inception, it was apparent that given the size, shape and the spans being attempted, the structure may be susceptible to human-induced vibrations.

It was initially unclear exactly which parameters within the context of the design could be used to control the vibrational performance of the structure. Therefore, from an early stage, it was important to develop a quick method for generating structural analysis models in order to perform a sensitivity analysis which would allow the design to progress rapidly. In this situation, the use of a sensitivity analysis was important, as the complexities and tight architectural constraints meant that it wasn't immediately obvious how various design changes would affect it. The sensitivity analysis allows the design engineer to develop a 'feel' for the structure, which makes communicating the key principles considerably easier.

Given the size and scale of this project, the primary challenge was to understand how the excitation of one area may affect another. As the full ramp is one long, continuous structure, this is not at first intuitively obvious: the excited point would seem to be the one to experience the largest response. An example of this would be a diving board: the end of the board from which the diver jumps also experiences the largest amplitude of vibration. However, for more complex structures, this relationship is more difficult to understand. It is also the case that the person walking (or running) up a staircase is not actually the one experiencing the vibration at the point they've excited as they've already moved on.

Due to the infancy of this type of analysis, there are no code provisions from which to design; however, it is now most common for a response factor to be specified. Response factors essentially convert the acceleration at a particular point to the level of comfort for a typical person. Relating this to the comfort level of a typical person on the structure inherently brings an element of probability into the analysis, since every person experiences vibrations differently. Nevertheless, the use of the response factor is considered more relevant and accurate than the historical method in which the likely frequency of excitation is compared with the structure's natural frequency, and where they are found to be similar the structure is modified to avoid the problem.

To conduct the analysis, the structure was divided into a series of steps, and each step excited at a range of frequencies that people typically walk or run at (considered to be in the range of 1.8–4.5 Hz for a staircase): this range led to around 2,700 load cases to analyse. However, while there were only 2,700 load cases to run, as the response of each step to the excitation of all the others was also required, it meant extracting close to 300,000 unique acceleration values (Figure 5).

3 Tacoma Narrows Bridge, Tacoma, Washington, USA, 1940.
The bridge collapsed shortly after opening due to divergence, a wind-induced dynamic phenomenon.

4 Foster + Partners, Millennium Bridge, London, UK, 2000.
The bridge was closed shortly after opening, due to excessive human-induced vibrations, and retrofitted with dampers.

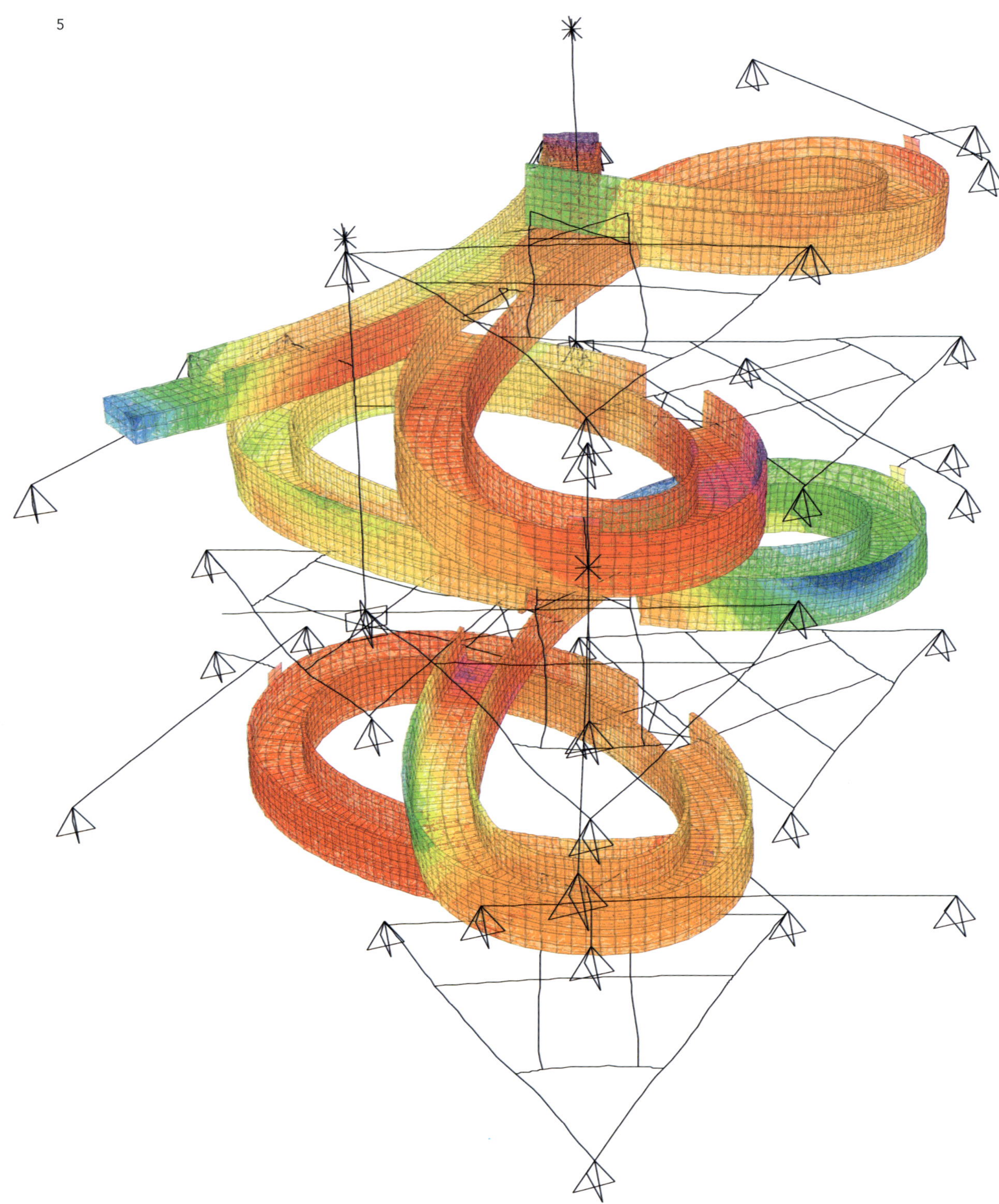

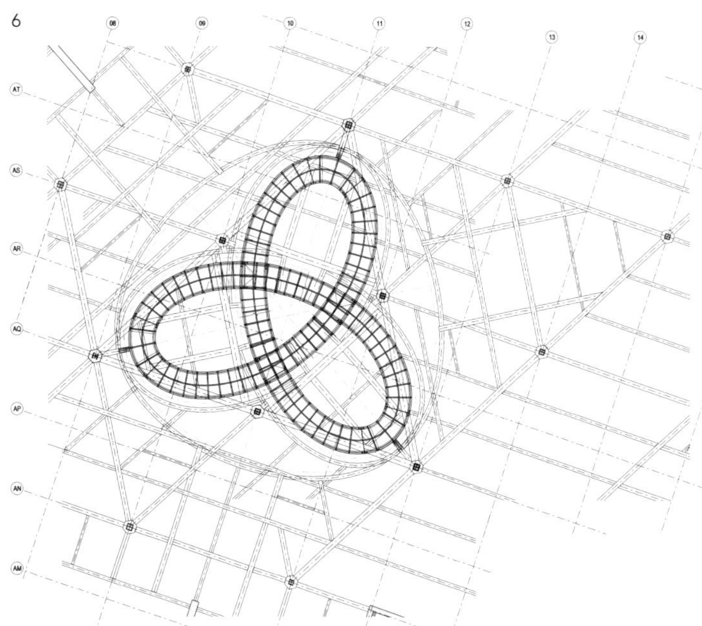

5 Foster + Partners with AKT II, Bloomberg headquarters, London, UK, 2017, analysis model of the Bloomberg stair.
The image helps to visualise the area of the structure excited at a particular frequency.

6 Foster + Partners with AKT II, Bloomberg headquarters, London, UK, 2017, hypotrochoid ramped staircase.
The continuous structure of the Bloomberg stair rises six storeys through the main atrium of the building.

Only through this broad-brush approach to the analysis was it possible to understand which areas of the structure were most susceptible to excitation. It led to the production of a considerable amount of data, which then had to be formatted so that the rest of the design team could understand the behaviour of the structure. The graph in Figure 7 shows how each step of the ramp responds to each of the others being excited. It would be expected that the maximum response of each step would occur when it is being excited. This can be seen in the graph as peaks occurring along the leading diagonal. However, the graph also shows significant contouring, which indicates it is not just the excitation of the step that causes a high response factor, but also of adjacent steps.

The challenge in this project was not only the complex and data-intensive analysis, but also the continual development of the fabrication methodology. Initially, to ensure continuity within the structure, it was designed as a single, fully welded, monocoque structure, but as the programme for construction changed, the contractor's preference moved to a segmented construction method where each flight would be broken into a number of pieces that would be bolted together on site. Coordination of this within the scope of the architectural constraints presented a significant challenge, along with ensuring the ramp conformed to the original design of a continuous structure top to bottom. Figure 8 shows the design work that was carried out to understand how the continuous structure could be appropriately divided into segments, and how these would come together on site to achieve the same continuity the design required.

Due to this close coordination work, the complex ramp was delivered on site within the original programme. The image in

7 Foster + Partners with AKT II, Bloomberg headquarters, London, UK, 2017, cut-away of ramp section.
This highlights the work that was carried out to develop the segmented construction methodology to aid the on-site construction.

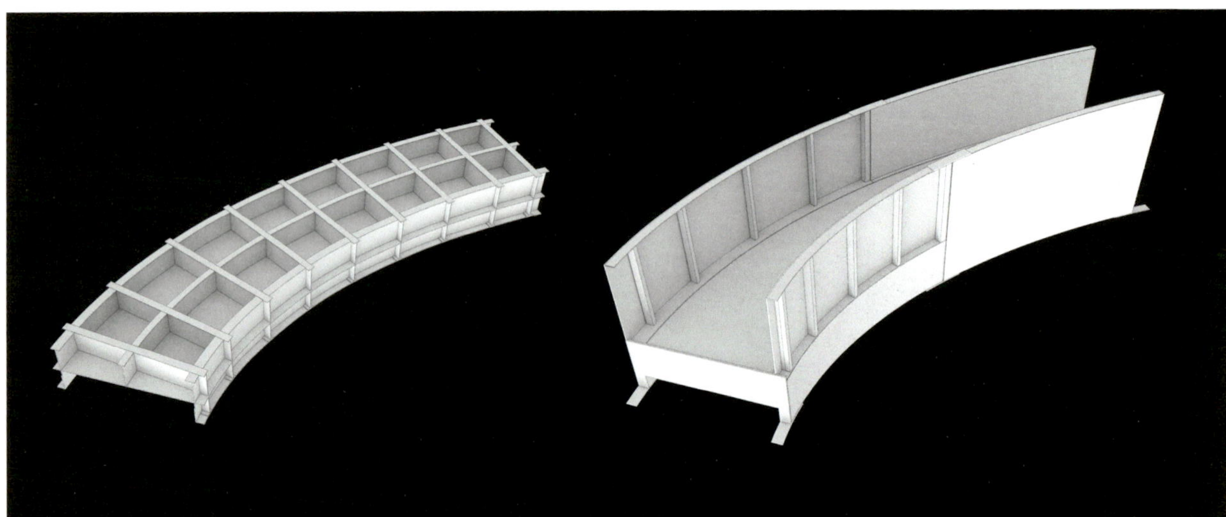

7

8

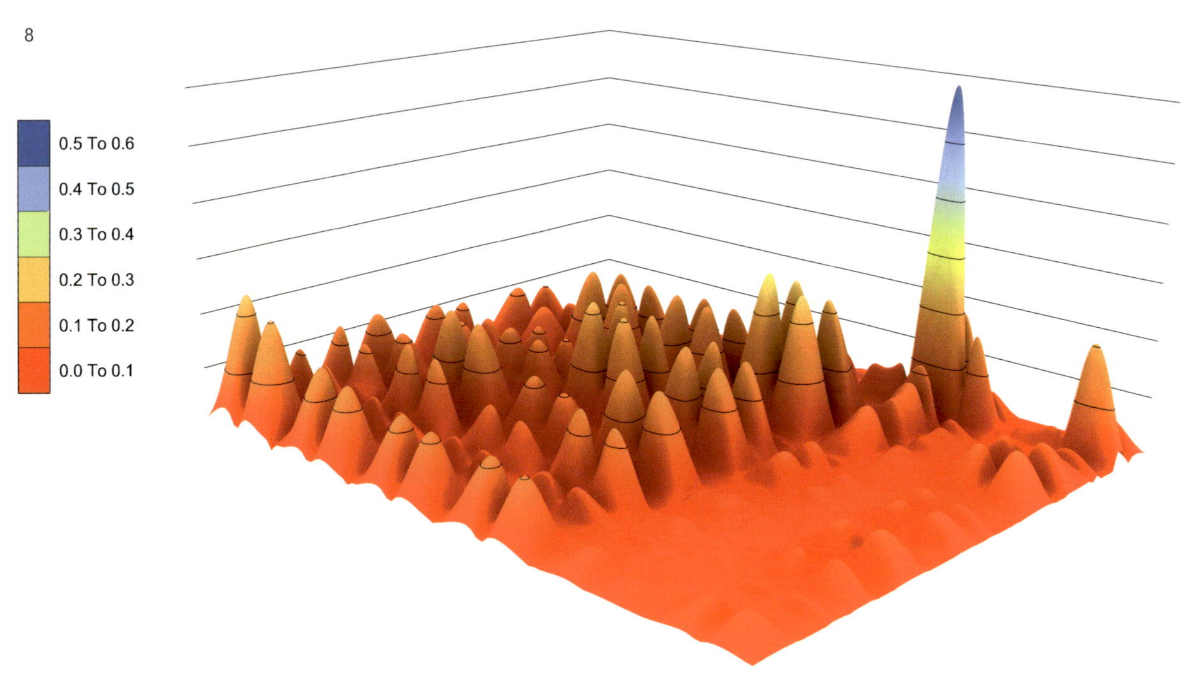

8 Foster + Partners with AKT II, Bloomberg headquarters, London, UK, 2017, response factor distribution for entire staircase.
The figure indicates how each step responds to each of the others being excited.

Figure 9 shows the ramp *in situ*, expressing its grace and elegance in the context of the rest of the building.

In projects where the primary concern is human-induced vibrations, the response factor approach is being used increasingly. This approach is interesting, as it is no longer based on the pure structural mechanics discussed previously, but introduces a probabilistic element to the design: human perception. To an extent, this could be considered a form of performance-based design, as it essentially allows the client to choose the level of performance they require their structure to meet. The more onerous the performance criteria, the more limited the scope of design or the greater the cost implications. The performance-based design approach was developed for the design of structures in high seismic regions where defining various performance criteria and levels can achieve greater efficiency in design, with the client understanding from the outset which performance level their structure has been designed for.

By responding quickly and analysing early iterations of the structure's design, it was shown that there was a significant cause for concern over the dynamic response caused by resonance due to human-induced vibrations exciting the structure at its natural frequency. This led to three cases being established for the design: normal occupancy; egress-induced vibrations such as during an emergency evacuation; and event loading, which considered the loads induced by people jumping during a concert. Using these three categories, the response factor for each type of loading could be relayed to the client. The worst case was shown to be that of a sizeable crowd jumping during a concert. By relating this to a performance level, the client was given the choice of how the design should be developed: either the attendance at concerts could be limited to a safe number or additional engineering solutions could be applied to reduce the response factor at an increased cost.

Given the slender nature of the structure, reducing the magnitude of the structural vibrations by changing the design was undesirable. It was therefore established that the only method for reducing the vibrations was to make use of tuned mass dampers. These are often used in tall buildings to relieve the problem of wind-induced oscillations – Taipei 101 for example (Figure 10) – as well as on buildings in seismic regions. Tuned mass dampers work by reducing the magnitude at a particular frequency they are tuned to. This makes them highly effective where a structure is excited at a specific frequency, but less so when the range of frequencies excited is more varied.

The effect of using tuned mass dampers is shown in the graph in Figure 11 showing how the magnitude of the response decreases for different levels of damping. Analysis from this project indicated that the use of tuned mass dampers could reduce the response factor by up to 200%, which allowed the structure to hit the performance criteria desired.

9 Foster + Partner with AKT II, Bloomberg headquarters, London, UK, 2017.
Site photograph showing the Bloomberg ramp spiralling up through the central atrium of the building.

RESPONSE FACTORS
VERTICAL
No damping
Damped
Damped
250
200
150
100
50
0
RESPONSE FACTOR
1.4
1.9
2.4
2.9
3.4
3.9
PACING FREQUENCY, f (Hz)

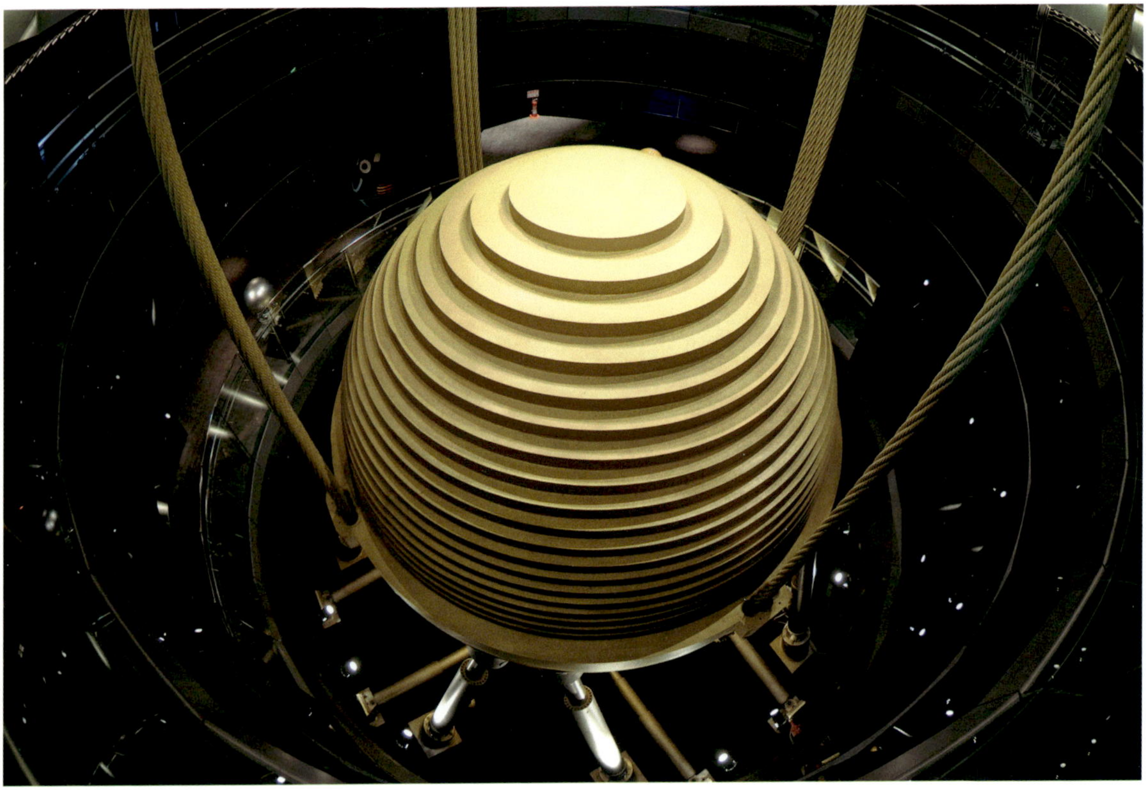

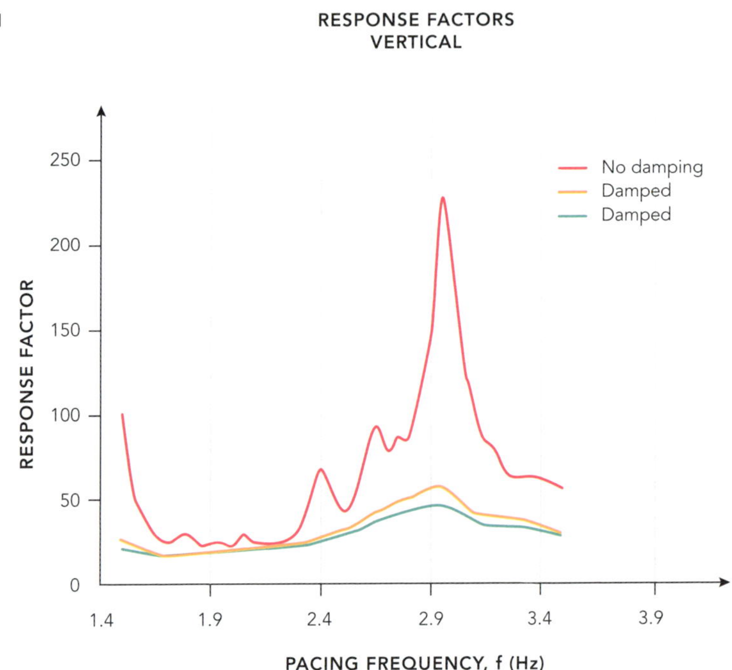

10

11

The examples described above show the ways in which dynamics, from the very outset, plays a vital role in how a project progresses. In both of these examples, and numerous others, it was not the static strength considerations that governed the engineering aspect of the design, but the dynamic response; failure to recognise these problems at an early stage would have had serious consequences at a later stage. To a large extent, the design engineer's ability to understand these issues is related to the exponential increase in computing power over the last forty years. Nowadays, engineers can test and iterate solutions which allow them to develop a much deeper understanding of how a structure is behaving. This improvement also makes them much more confident in communicating to the rest of the design team and taking a proactive lead in the design.

Dynamics though, is increasingly finding its way into other aspects of design engineering; it is no longer used only for determining whether the external forces applied to a structure will cause undue issues with vibration. Since the 1960s and 1970s, dynamics has been used as a method of form-finding known as dynamic relaxation. Initially this computer-based technique for form-finding was no match for the results and analysis that could be performed on scaled models, as the likes of Heinz Isler became extremely adept at constructing and interpreting. However, over time, as the role of computers in design has increased, this method has become more and more common, and nowadays would be used in preference to scale models. Dynamics is not just relevant on the macro scale, however; research to improve the modelling of concrete has looked at treating it as a series of particles dynamically connected to one another. This method, termed 'peridynamics', overcomes many of the issues associated with continuum mechanics, and offers the potential of more accurately modelling a complex material such as concrete.

10 The tuned mass damper used in Taipei 101, Taiwan, 2004.
The damper reduces excessive deflections of the 508 m structure during a typhoon.

11 AKT II, diagram highlighting the dramatic reduction in the response factor through the use of a tuned mass damper.

TEXT

IMAGES

11 FORCES OF NATURE

JEROEN JANSSEN AND
MARC ZANCHETTA

Wind, this wonderful, plentiful and natural resource, has moulded the face of our planet over the course of aeons. It has an equally important role in shaping the built environment and our perception within it; providing fresh air, removing pollutants, carrying sounds, cooling, chilling and providing a valuable source of renewable energy.

But wind is lazy and unpredictable. How do we model its passage across our built landscape and, most importantly, how do we design with it? How should we arrange the urban morphology to capture its beneficial effects, while shielding ourselves from the nuisance of gusts and the dangers of gales? How do we want our buildings, structures and landscapes to respond to the wind, and how do we test our designs?

Predicting the turbulent motion of wind flow remains a state-of-the-art problem as old as time itself.

1

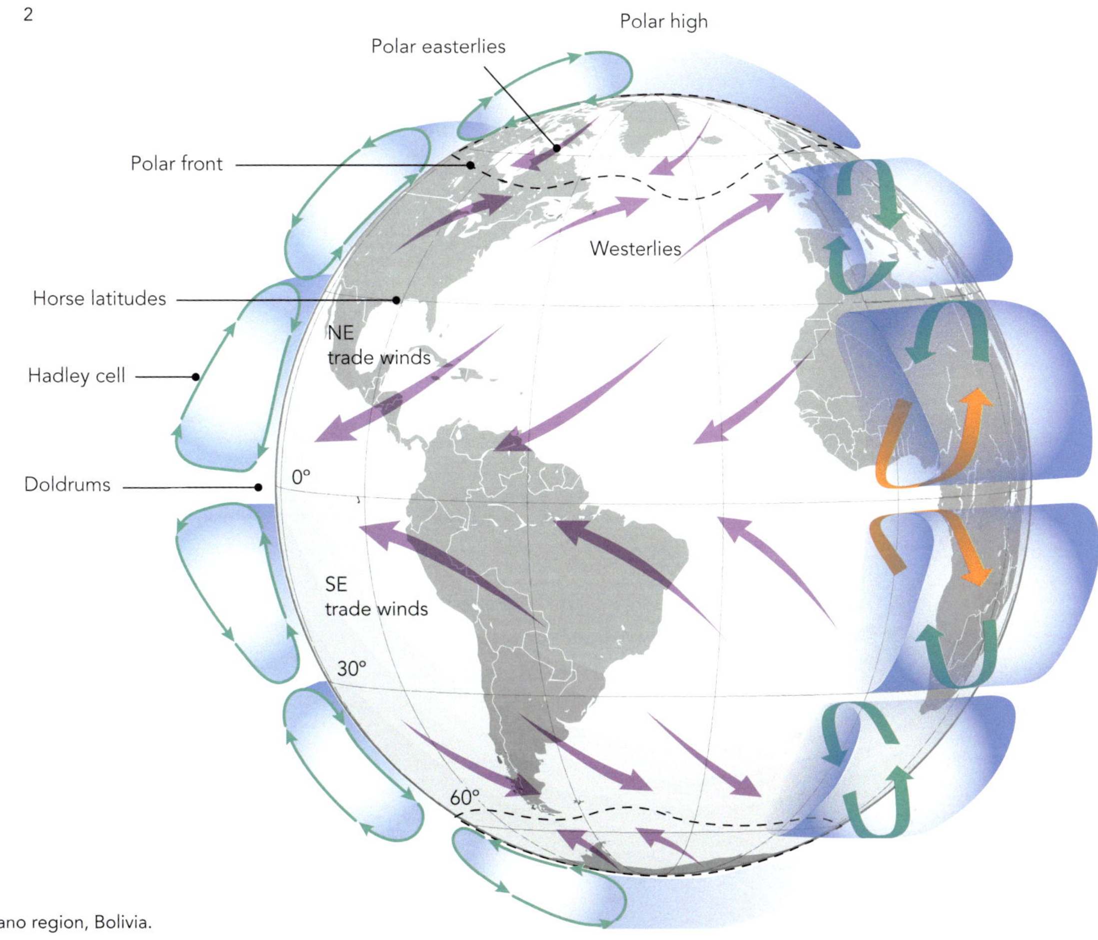

1 Altiplano region, Bolivia.
Landscape with solitary rocks sculpted by the wind.

2 AKT II, large movement of air masses creates global wind patterns driving meteorological variations in wind climates.

However complex the task, there are tools available to test designs, enabling design engineers to reach for the skies with their skyscrapers, while ensuring a pleasant and enjoyable environment for those inhabiting the ground plane. The wind tunnel remains the workhorse of industrial aerodynamics, providing invaluable answers to the toughest questions in the most demanding of situations. The recent advent of the numerical wind tunnel has changed how wind movement across the built environment is explored, and although its use has become prevalent, the question remains whether it can be used to make fit-for-purpose predictions to support designs.

Potential with wind is unlimited; as it works in unison with vegetation to form the lungs of the planet, its motions can be captured to provide cooling breezes in summer, natural ventilation in energy-efficient buildings, wind shelter in winter and renewable energy throughout the year.

Designing with wind also leads to conflicting goals: design engineers may wish to provide shelter in the cold winter months,

3

Regional wind
Vz
Vz
Vz
Urban
boundary layer
Urban
canopy
Urban
'plume'
Rural
boundary Layer
RURAL
SUBURBAN
URBAN
SUBURBAN
RURAL

4

Tall building:
flow mainly along the side
Intermediate or
transitional type

Wide building:
flow mainly over the roof

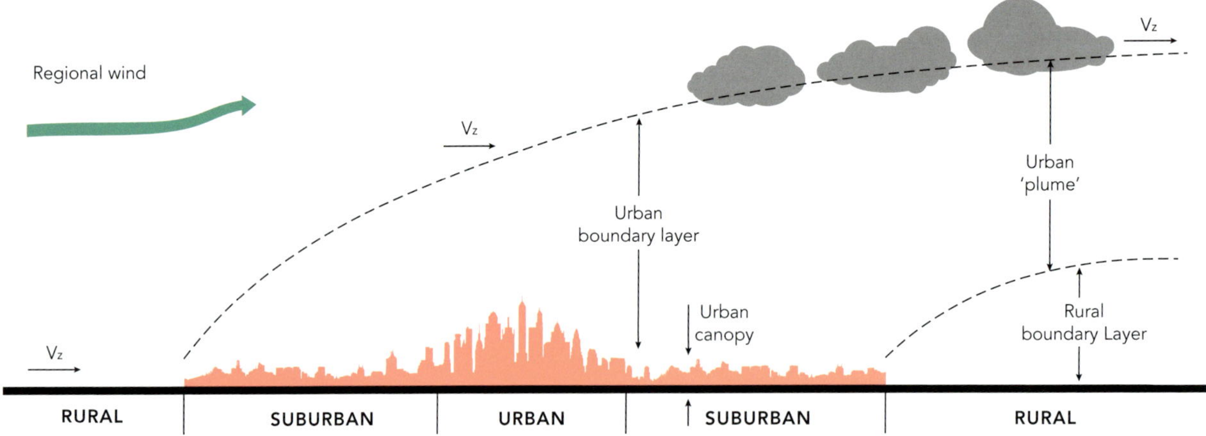

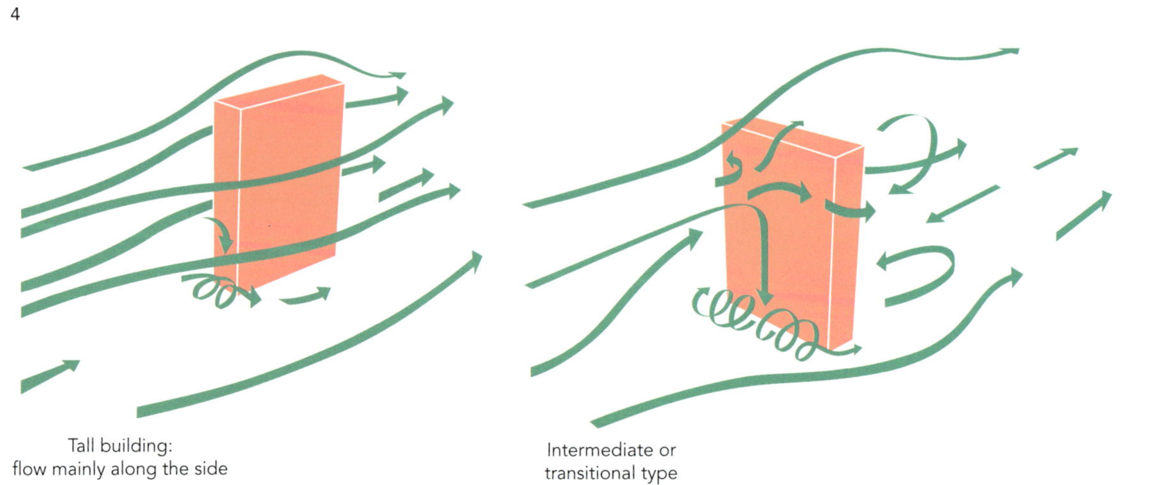

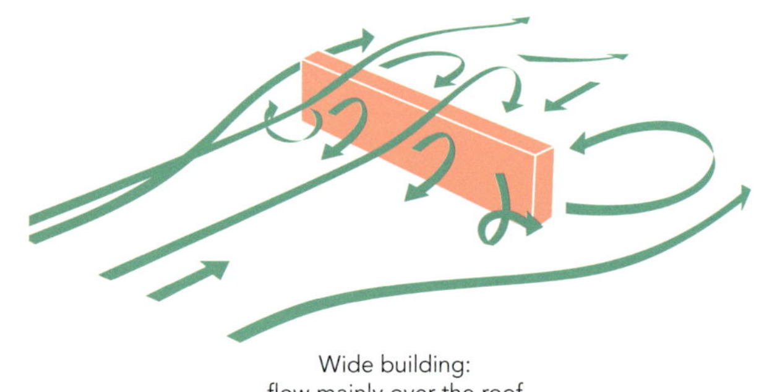

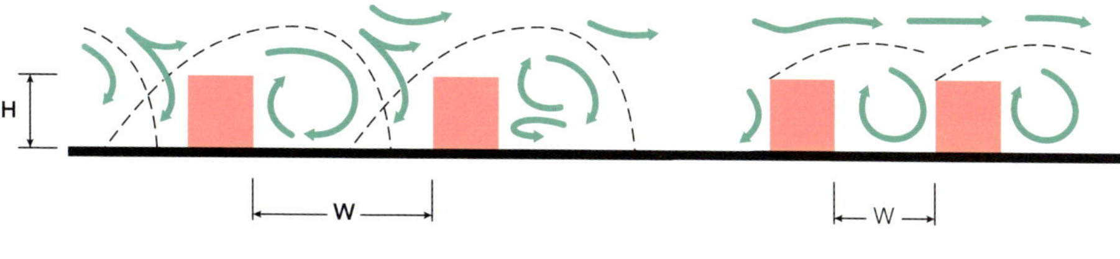

yet enhance beneficial breezes in summer, or provide large pressure differences across facades to drive natural ventilation, yet minimise wind loading on towers.

The problem is further compounded by the fact that this dramatic climatic force presents seasonal and diurnal variations, and its effects are often compounded by temperature and rain.

Wind is a result of the large-scale movement of air masses around the globe, and is driven by pressure differences resulting from differential heating between the equator and the poles. This complex motion is further compounded by the Coriolis effect, created by the earth's rotation.

Once a matter of life and death, many of these prevailing patterns have long been observed and harnessed to facilitate journeys across continents, to grind grain, or dry hops. Avoiding the doldrums or harnessing the trade winds, surfing the jet stream or capturing land and sea breezes – these have been familiar endeavours to many. Winds vary from point to point on the globe and also in time, with speed and direction responding to the macroscopic weather systems, and to the local topology and environment.

At height, winds are free to travel unimpeded, but at ground level the roughness of the earth's surface retards the motion, and the atmospheric boundary layer is formed, with the characteristic turbulence and chaotic motion that is experienced as day-to-day wind. This effect is most pronounced over cities: the wind speed at the spire of St Paul's is the same as the speed at Heathrow airport.

3 AKT II, schematic representation of the urban atmosphere, showing a build-up of the Urban Boundary Layer above the build environment.

4 AKT II, typical wind patterns around solitary buildings following different orientations to the wind direction.

5 AKT II, flow regimes associated with different urban geometries and proximity between blocks.

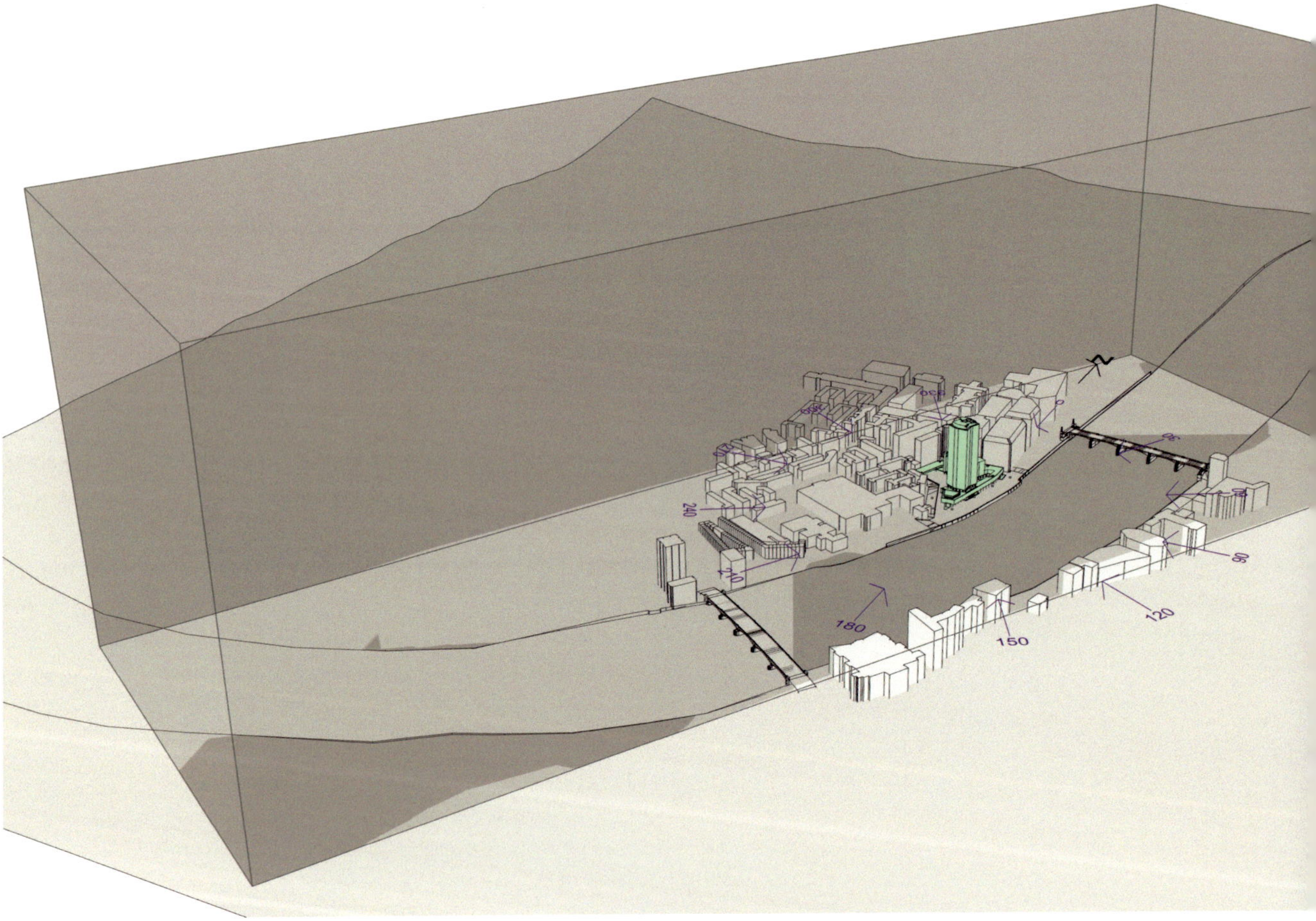

6

7

But this shear in velocity over cities, with the large defect in velocity at ground level, carries hidden dangers with it: downdraughts and jets are the scourge of many poorly planned developments. Canopies and breaks can be introduced later to mitigate the effects of a poorly laid-out masterplan, but the wind engineer knows that each individual massing must be carefully tuned in its interaction with the wind during the earliest design phases of a project.

Past experience teaches much about the dos and don'ts in urban layout, and the planning process requires careful consideration to ensure that the wind environment will be suitable for the intended use. Setting aside the complexity of wind motion, designing with wind, like any other aspect of building design, requires an ambition, a brief and an assessment approach.

Designing with wind means creating and exploiting differences in velocity and air pressure to harness and steer its motions across the site. To make these phenomena meaningful and measurable for design, they can be described in degrees of comfort.

Comfort is a physiological and psychological state expressing satisfaction with the environment. Not one condition exists that will satisfy all occupants. The importance of a place's environmental qualities, such as temperature, noise, acoustics, smell, light, heat and wind, must be considered alongside the occupant's desire, or lack of it, to remain in the place. While the psychological state remains the primary concern of the architect, the environmental engineer can support the design of the physiological part. In so doing, the conflict between those who take a purely physiological view of comfort and those who take a purely psychological one can be resolved, enabling the development of a brief that responds to the season and time of day for all user groups within a space.

OUTDOOR MICROCLIMATE COMFORT
The wind engineer will often consider wind in isolation, answering the questions: how windy is it here and how often? Is it safe and is it comfortable? To answer the first question, the likelihood of dangers arising from the high winds and the severity of the possible impact must be addressed. To answer the second, the intended use of the space must be established: a space that will frequently be too windy for a picnic might provide a refreshing route for walking and running. The problem is further compounded by the fact that a person's tolerance for wind will be modified by their expectations: an invigorating breeze at the seafront will be regarded as a tonic, while a gusty environment at the base of a tower will only ever be a nuisance.

Computational power is such now that this question can be answered directly with digital simulations; the need for a skilled wind engineer, however, remains as important as ever since the risk of non-physical results looms large.

6 AKT II, Millbank Tower redevelopment, London, UK.
Overview of the numerical wind tunnel set-up showing the computational volume with the tower and surrounding buildings modelled for an accurate simulation of wind flow.

7 AKT II, Millbank Tower redevelopment, London, UK.
Visualisation of simulated streamlines around the development for wind blowing from ENE (60°) direction as a visual output of the Bioclimatic Comfort Toolkit.

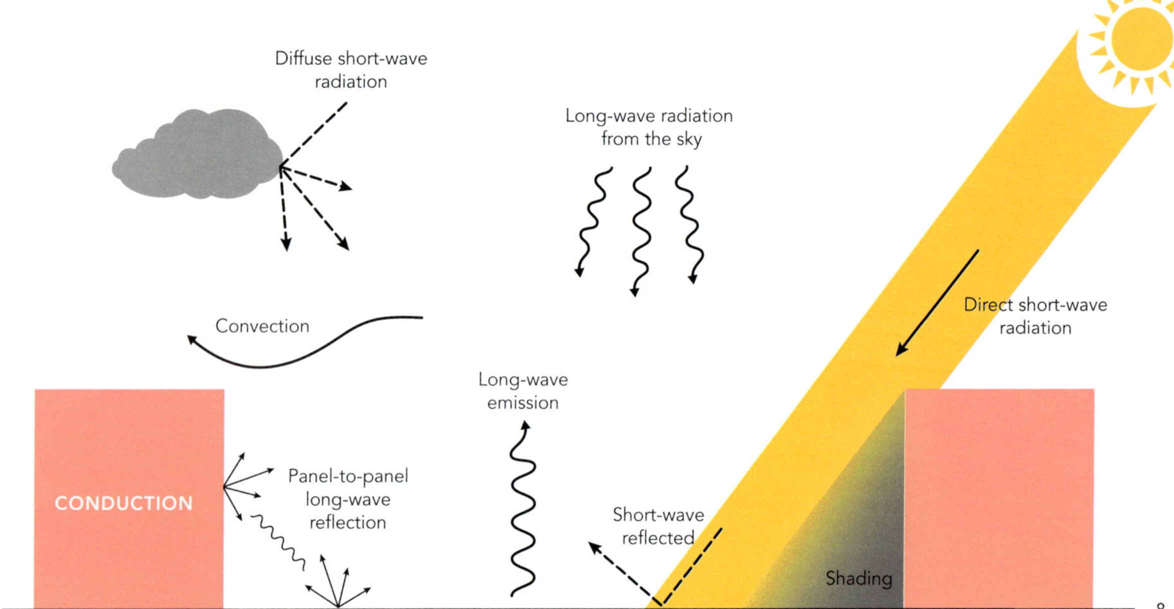

Accurate simulations can be set up for the full range of wind directions, and with historical measured wind data these simulations can be mapped to predict the wind regime over the course of a year. The dataset will describe the probability of certain wind speeds occurring, as well as their direction and duration. With this statistical view of the wind climate around a site, we can define the acceptable occupation of a certain space, and come to a decision as to whether it is comfortable for this use or not.

The modelling fidelity can be increased beyond the simple wind tunnel with the aim of predicting the full environmental behaviour of the external habitat under the combined influence of not just the wind, but also the sun, heat, clouds, buildings and shade. Just as the building services engineer can model the sensation of comfort within an enclosed space by considering air movement, air temperature, thermal mass, shade, light and solar radiation, the numerical wind tunnel can be expanded to examine the complete thermal environment of the urban space.

To do so, the motion of the sun must be tracked, hour by hour and day by day. How shadows move across the space, how solar radiation is received, absorbed and reflected (Figure 8), and how heat is radiated back to the sky must all be predicted. The following must also be modelled: how heat is transferred between surfaces and stored in depth, and – using the previously described tools – how wind cools (or heats) surfaces. In short, an hour-by-hour prediction of the evolution of the main environmental qualities in a space must be built. Once this is done, the outdoor environment can be compared to the outdoor microclimate comfort brief.

BIOCLIMATIC COMFORT TOOLKIT

As part of recent research and development, AKT II, together with Tyréns UK and GasDynamics, has undertaken the necessary analytical and numerical work to develop a simulation tool: the Bioclimatic Comfort Toolkit. It is a world-leading tool unique to the market, and able to inform design in real-time. This application connects the 3D CAD environment with a physiological and physical model, simulating the effects that the climatic inputs of a site have on its materials, geometry and local urban microclimate. The toolkit then evaluates the results of the simulation and visualises the thermal comfort levels in a CAD environment, providing an easy-to-use interface for the designer.

To predict the degree of comfort, the way the site tempers the local climate needs to be looked at or, to put it more simply, the microclimate needs to be predicted. This involves building an hour-by-hour simulation of the effects of air temperature, solar radiation, relative humidity, wind velocity, direction and the build-up of surface temperatures. The climatic inputs can be read from a weather file, which is generally composed of thirty to fifty years of measured data. The geometric properties are modelled in the 3D CAD environment and the material properties assigned.

Before running the simulation, information about the geometry of the site is gathered. With the sun path in hand, and the boundary surfaces of the site divided into panels, a shadow map, a shape-factor map and a sky-view factor map are then built. These provide the basic information necessary for the simulation in every time step and are only computed once, before the simulation starts.

The shadow map determines the hourly shadow patterns for the site. To do this, a tool has been embedded that creates a fish-eye render for every panel in question, looking upwards from its centre point. That image is analysed with the hourly position of the sun overlaid; it determines for every hour whether the panel is in the shade or not (Figure 9).

The shape factor is a geometrical relation or ratio of how much each panel sees of all the other panels. This is mathematically defined and dependent on the area of both panels, the distance between them and the angle between the normal. The information is stored in a large matrix, required for the long-wave radiation simulation between the panels.

Lastly, the sky-view factor describes how much of the sky is visible from every panel. This map uses a single ratio for every panel and can be visualised as a gradient from black to white. This ratio is defined as the remaining area after summation of the shape factor, projected on a unit sphere (Figure 10).

With these global inputs in place, the simulation marches forward in time to predict the microclimate with time steps of one hour. The computation takes several independent radiation inputs, starting with the direct solar radiation, or the direct influence of

9
10
11
12

sunbeams. The intensity is combined with the information in the shadow map, turning the input on or off for specific panels and specific hours in the simulation. Other inputs are the indirect solar radiation, or the short-wave reflection from the sky, and the long-wave radiation from the sky.

Following the simulation, the heating or cooling of surfaces by convection is examined, using an independent numerical wind-tunnel simulation that describes accurate local wind climates around the geometry on site. The resulting wind velocities and directions are then compared to the orientation and inclination of the surface panels (Figure 11).

Long-wave radiation reflected by the geometry is then assessed. This includes the diffuse reflection of incoming radiation to and from all the panels, with the reflectivity of the surfaces defined in the geometric material set-up of the scene and the amount of reflection computed with the help of the shape factor.

After looking at all these external influences, conduction through the surfaces is considered. Heat transfer is dependent on the heat flux coming in, material properties, layer thickness, material density and the internal temperatures in the previous time step. A total heat flux for each panel for a time step is determined, and once the surface temperatures are updated, temperatures are plotted as a colour gradient in the 3D CAD environment and the simulation continues to the next time step.

An analysis plane at head height that cuts through the urban domain can then be created to visualise the thermal comfort of a space. For every sample point, the Universal Thermal Comfort Index (UTCI) equivalent temperature can be computed and plotted back on a colour gradient to show how successfully the space performs (Figures 12 and 13). With these results, the shape of buildings can then be optimised and urban interventions compared.

Predicting the wind behaviour around a site by simulation has always been difficult. In light of this, a direct link to OpenFOAM, an open source computational fluid dynamics (CFD) package, has been integrated into the toolkit. As simulations in CFD are very intensive to run, but essential for an accurate assessment of the microclimate and comfort, it has been included in the toolkit as an independent input into the full thermal simulation.

The user interface of the toolkit prepares the required set-up for the digital wind tunnel and geometry. The volumetric mesh for the simulation is created automatically and simulations are then run. These computations are usually run for the full spectrum of wind directions, providing a complete dataset of velocity, directions and pressure. They can then be related to the wind on the specific site, statistically matching the probability of the wind blowing in a certain direction with a certain velocity. The results can be used not only as an input for the convection simulation, but also for a separate pedestrian wind-comfort assessment.

9 AKT II, Centre Point, London, UK.
Visualisation of shadow map for 29 April 2015, 13.00h, as a visual output of the Bioclimatic Comfort Toolkit.

10 AKT II, Centre Point, London, UK.
Visualisation of sky-view factor as a visual output of the Bioclimatic Comfort Toolkit. Clearly visible is the gradient from wider streets to narrow and dark alleys.

11 AKT II, Centre Point, London, UK.
Visualisation of simulated surface temperatures for 29 April 2015, 13.00h, as a visual output of the Bioclimatic Comfort Toolkit.

12 AKT II, Centre Point, London, UK.
Visualisation of simulated thermal comfort (UTCI equivalent temperature) for 29 April 2015, 13.00h, as a visual output of the Bioclimatic Comfort Toolkit.

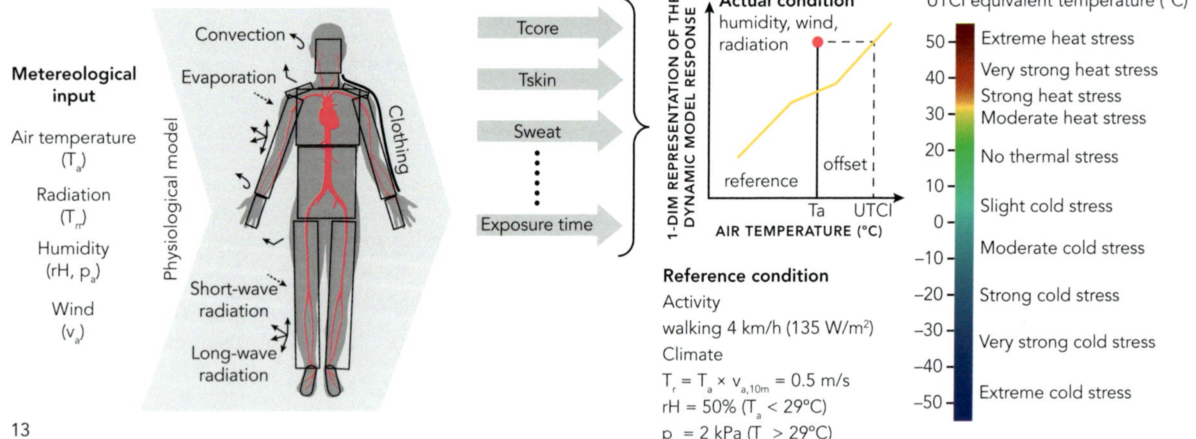

13

INTERROGATING THE SIMULATION

The Bioclimatic Comfort Toolkit offers the possibility of asking many questions, the nature of which will depend on the task in hand. Does the new development reduce the amount of light enjoyed by a critical receptor? Do the habitable rooms of the proposed massing experience sufficient daylight? What sort of shading best suits a particular facade elevation? Are the wind speeds too high near the proposed entrances? Can we use vegetation to shade our preferred route and enhance wind movement? Where and how often will sunshine be enjoyed in a courtyard? How is heat conducted in depth? But one question encompasses all of the above: how comfortable is the external environment?

Therefore, the proposed solution must be interrogated to discover how the space will be perceived. To do so, a series of control points within the domain are located and the wind, sun and radiation allowed to act out their effects. A variety of comfort engines are being used in different fields of design and engineering, many of them geared towards the internal comfort of a space, such as the physiological equivalent temperature (PET) approach that links physiological comfort with thermal stress associated with heat losses and gains, or the actual sensation vote (ASV) method, which has correlated a sensation vote with a broad measurement of the climatic parameters. Adopted in the toolkit is the UTCI, developed by the International Society of Biometeorology, taking into account all the possible combinations of physiological climatic inputs together with the Fiala body-model, a complex model simulating human heat transfer inside the body and at its surface.

THE IMMEDIATE FUTURE

Recent events with newly built high-rise buildings in London have highlighted the importance of specular reflection and focal effects from glazed facades. An extension of the tool to cater for this, as well as the more traditional problems of glare and dazzle, seems a logical next step. This may be best achieved with some principal rules covering first-bounce specular reflection.

As projects continue to be designed further afield, comfort models must expand to cater for regional differences and adaptive behaviour, as well as seasonal and diurnal expectations and responses.

The impact of all bioclimatic interventions in the designer's armoury must be adequately predicted: water pools, localised shading, evaporative coolers and radiant heaters. Most importantly, all predictions must continue to be validated.

THE TASK OF VALIDATION

A simulation, be it numerical or experimental, is only a model of reality, often capturing only a fraction of the physical or geometrical complexities of the real world. Such short-cuts are necessary if desktop models that can simulate events are to be built in the future. Validation and verification are the processes by which design engineers understand the predictive power of the tools and understand whether they are fit for purpose. While the individual tools may be validated in isolation, validating the entire tool set remains more arduous and relies on rigorous academic studies. When discrepancies exist, it is important to question whether they invalidate the work or simply serve as a reminder that an engineer's predictive ability is constrained by the physical and geometrical complexity of the project.

Equally, and perhaps most importantly, we need to know if we are asking the right questions of our simulations: are we looking for differences that cannot be resolved, or are these trends not being spotted because of the way data is interrogated, or is the implemented physics engine not sufficiently complex for the model to capture the differences? A black-box approach to numerical modelling will never suffice.

There are few designs that will test all extremities of the microclimatic effects. Design engineers must always be aware that, when designing a park in Abu Dhabi, for instance, it will be impossible to recreate London's Hyde Park on a sunny day; it will be impossible to make the space comfortable for everyone, for the whole year. It may, however, be possible to extend the use of the space, effectively extending the usable season by a number of weeks or months. In climates as different as these, the comfort brief needs to be developed thoroughly, and it should describe the intended use of different spaces and the critical duration for these to be pleasant. In short, the Bioclimatic Comfort Toolkit reinforces, rather than replaces, the need for the designer, and will simply arm him or her with more data to support the design process.

Predicting in a coupled fashion, or better dynamic fashion, the wind action on a structure and the subsequent structural response to the wind, remains the primary objective. As innovation continues, it is paramount that structural design responds to the

13 AKT II, Universal Thermal Comfort Index (UTCI).
Overview of the calculation model to represent a comparable value for thermal comfort, as implemented in the Bioclimatic Comfort Toolkit.

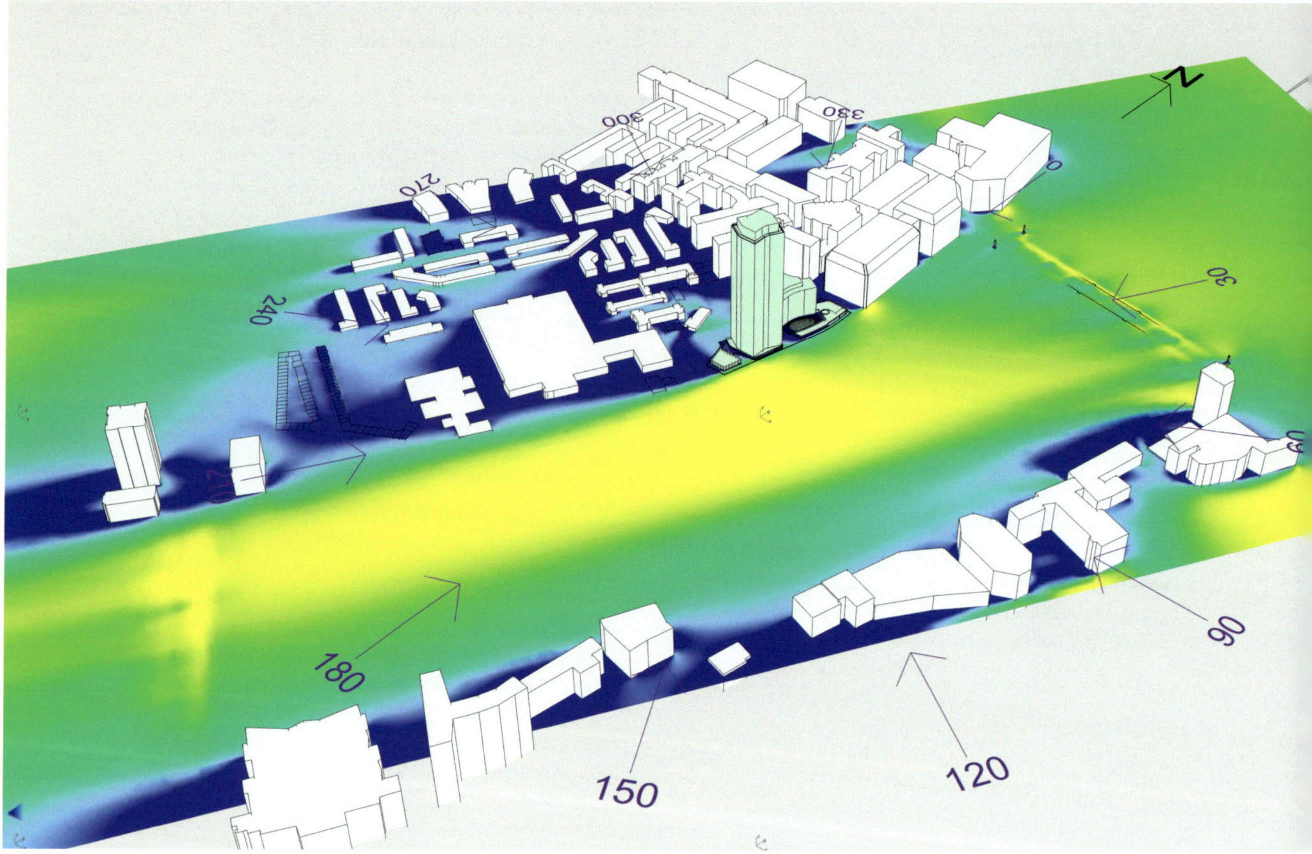

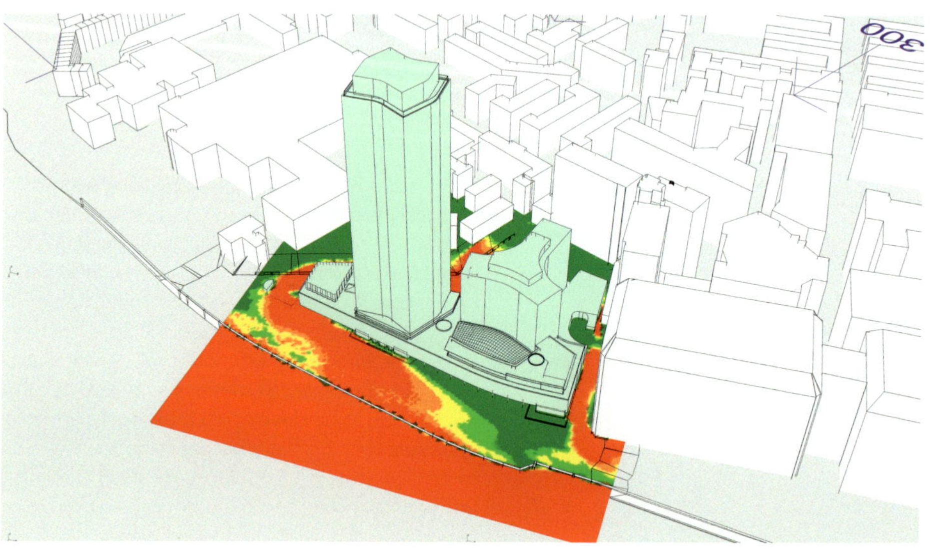

14

15

dynamic effects induced by wind and develops its ability to predict and then mitigate the impact.

Aeroelasticity is the name of the game, requiring an extensive and expensive computational deck. Advances in numerical wind engineering endeavour to capture flow physics in ever greater detail, and the topic remains a state-of-the-art problem for the academic community. The intractable nature of turbulence continues to overwhelm even the greatest of supercomputers. However, many problems can be resolved in a numerical wind tunnel, and the ability to explore wind loading helps us to better understand the effects of wind action on complex geometries early in the design process, which is of immense importance. Design engineers today look forward to sharpening pencils once more, turning to the computer to support dialogues with architects.

14 AKT II, Millbank Tower redevelopment, London, UK.
Visualisation of simulated wind velocity as a colour gradient around the development, for wind blowing from NNE (30°) direction, as a visual output of the Bioclimatic Comfort Toolkit.

15 AKT II, Millbank Tower redevelopment, London, UK.
Visualisation of pedestrian wind comfort around the development as a visual output of the Bioclimatic Comfort Toolkit.

12 STRUCTURAL SKINS

MARCO CERINI

Design engineers today have more choices than ever. Modern buildings, especially those with more than two storeys, are often designed and built as frames, which are then clad with the desired facade panels, whether glazing, brick walls or otherwise. Concealed or not, the skeleton of the building remains a frame of columns, beams and floor plates.

Sometimes a different approach may prove more efficient, at least from a support perspective. Given that the envelope is there anyway, getting it to share the supporting function, or even perform it entirely, can yield considerable efficiency. This is not an entirely new idea in the building world; before the advent of reinforced concrete and structural steel, the building envelope was also the principal bearing element, for example when we think of the bearing brick-wall cathedrals of the Renaissance.

What we are now experiencing in architecture can be compared to the era that saw development of the first aircrafts. The

1 Typical construction of first aircrafts: Bristol F.2B fighter bomber (designed in 1916).

2 Slim Concorde supersonic transport, second prototype, assembled at Filton, just outside Bristol, UK.
The interior has tunnel-like appearance.
This prototype is concerned with the British elements of the joint Anglo-French programme. Another is under construction in France. Date photographed: 23 June 1967.

Wright brothers conceived their initial gliders and the very first powered heavier-than-air craft as a lightweight biplane timber frame, strengthened by bracing wires and covered in fabric. Early developments followed exactly the same idea until engine performance improved to the point where fabric skin could no longer efficiently withstand the increased speed and consequent air pressures. At the same time, early cotton fabrics were not proving very durable. Initially the fabric skin was replaced by a thin aluminium film. With the thickening of the aluminium cladding to sustain higher pressures, the skin could be utilised to work together with the frame and share the carrying capacity of the overall structure, thus allowing essential weight savings.

Eventually aircraft, in most cases, have become skins stiffened by ribs, spars and stringers (Figure 2). The monocoque construction has been born; the frame replaced by ribs that work together with the skin, the two stiffening each other in a mutually beneficial combination.

TRADITIONAL CONSTRUCTION OF BUILDINGS
Typical of the classical and modern periods were compression-type buildings, that is, those with brick-wall and unreinforced-concrete construction, with extensive use of shear walls, arches, vaults, buttresses and flying buttresses.

Since the advent and spread of reinforced concrete and structural steel, the frame typology has perhaps become the more typical form of construction. In the construction of our cities, concrete or steel frames, often linked to a reinforced-concrete core providing lateral stability, are a common sight. Framed construction is

3 Zaha Hadid Architects with
AKT II, Middle East Centre,
St Antony's College, Oxford,
UK, 2014.
Completed building.

4 Zaha Hadid Architects with
AKT II, Middle East Centre,
St Antony's College, Oxford,
UK, 2014.
Timber frame.

simple and convenient, allowing the supporting function to remain separate from the envelope function. The envelope can be designed, produced and installed almost independently of the structural frame, bringing benefits to the management and programming of the works.

More recently, we have seen that the frame typology is flexible and can adapt to complex geometries, as demonstrated in the two projects shown in Figures 3 to 6. Zaha Hadid's Middle East Centre at the University of Oxford is supported on a timber frame made of planar subframes. The 3D complexity of the cladding is here decomposed into a series of bi-dimensional geometries linked to one another at key nodes.

In a similar fashion, the smooth shiny facade of the Birmingham New Street refurbishment hides a complex steel frame that closely mimics the cladding shape. The result is a combination of 2D and 3D frames which support and shape the cladding and bridge it to the station's original concrete structure by spanning directly between the station columns. For example, the frame in Figure 6 is effectively a single space frame which spans 22 m from bottom right column support to top left roof support. It provides the supporting member for the cladding mullions by following the final shape within a given tolerance, and features a combined bending and torsion truss to support the cantilevering panels.

In these examples, the cladding doesn't contribute to the supporting function and is considered completely independent.

For all the advantages of this now common way of constructing, layers are often built up in order to mount one functional element on the other: this creates stratification, redundancy, space and weight 'waste'. In these situations, a different approach could be adopted, one in which the cladding contributes to the carrying capacity, especially since these examples do not feature any openings. Such an approach has its own complications, since the manufacture of doubly curved structural panels is not commonly available within the building industry.

STRUCTURAL ENVELOPES

New fabrication techniques now make it possible to conceive buildings where the envelope and structure coincide, much like the airframes of modern aircraft.

Buildings, however, are not constrained by the weight limitations of aircraft, and design architecture and its creative drive can give rise to much more varied and irregular geometries: the shape is only limited by one's imagination, the physically possible, the capacity to fabricate what is conceived and, often, the cost.

In new techniques, the typical prefabricated products of the building industry, such as rolled-steel sections, are replaced by simple flat metal sheets, which are then cut, worked, shaped and connected to suit. Figure 7 illustrates a stage of this process with

5 AZPML with AKT II, Birmingham New Street
station, Birmingham, UK, 2015.
Completed facade.

6 AZPML with AKT II, Birmingham New Street
station, Birmingham, UK, 2015.
A portion of the facade's steel frame.

6

7 CRAB with AKT II, drawing studio, Arts University Bournemouth, Poole, Dorset, UK, 2015.
Fabrication stages of a monocoque structure.

8 CRAB with AKT II, drawing studio, Arts University Bournemouth, Poole, Dorset, UK, 2015.
Completed building.

9 CRAB with AKT II, drawing studio, Arts University Bournemouth, Poole, Dorset, UK, 2015.
Surface definition and imperfections.

steel: laser-cutting flat sheets, shaping and forming them to the desired curvature, assembling them in the required arrangement, welding the sheets together. With these techniques, metal sheets become an extremely versatile semi-finished product that can be transformed into almost any geometry.

When compared with other structural products, metals are also good candidates for external enclosure as they are inherently waterproof. Carbon steel is subject to corrosion when exposed to the open atmosphere, but effective protective measures are available (although maintenance is normally required). Some metals, however, are almost completely immune to corrosion in several common atmospheric environments, as is the case with stainless steel, weathering steel and aluminium. In addition, the joining of metal parts by welding allows effective structural connectivity and enables the creation of a continuous and impervious envelope.

DRAWING STUDIO AT THE ARTS UNIVERSITY BOURNEMOUTH

An example of this new technique in practice is that of the recently completed drawing studio for the Arts University Bournemouth (Figure 8). The architectural design of the studio is by Cook Robotham Architectural Bureau (CRAB), founded in 2006 by Peter Cook (a founding member of the '60s Archigram group) and Gavin Robotham. A set of openings of different sizes, positions and orientations would create different light conditions within to suit various drawing environments. Besides satisfying the functional requirements, the desire of Peter Cook, whose own talent was developed at the Arts University Bournemouth as a young student, was to 'place there something intriguing among the trees'.

SURFACE GEOMETRY

This intriguing object was to take the form of an organic, smooth shape, opening up for the various windows and entrances. A large oval window is placed at the north end and is possibly the main feature of the building shape; another wave-shaped window is in the middle of the roof, also facing north. On the east side at floor level, a long narrow window draws light from beneath a bench. Two entrances are located in lateral pods stemming smoothly from the main building body. It is not easy to merge all these elements into a single smooth shape with the drawing tools commonly available in CAD programs.

An example of this attempt is shown in Figure 9. The surface is built from a set of subsurfaces which are generated from edge lines defined at key locations. The subsurfaces are generated using an interpolation tool available in commercial software. This is very much a manual process that has to be repeated every time the geometry needs to be updated during the design development. Figure 9 highlights some of the issues that are likely to be encountered. Some CAD programs have tools to generate complex surfaces from an incomplete boundary; if used in this case, they may create gaps. Even when boundaries are carefully defined, two contiguous surfaces are not guaranteed to

10

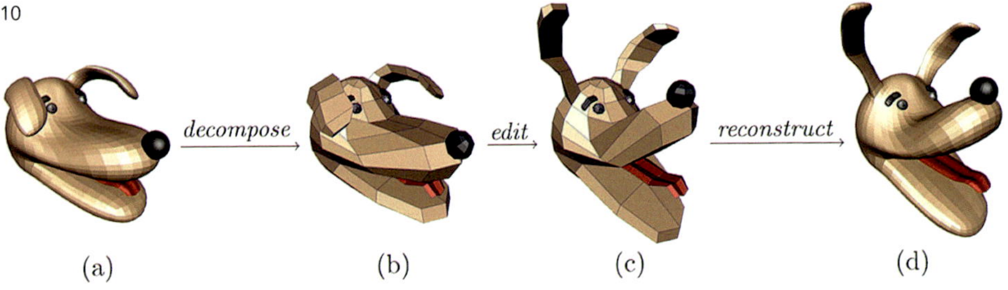
decompose
edit
reconstruct
(a)
(b)
(c)
(d)

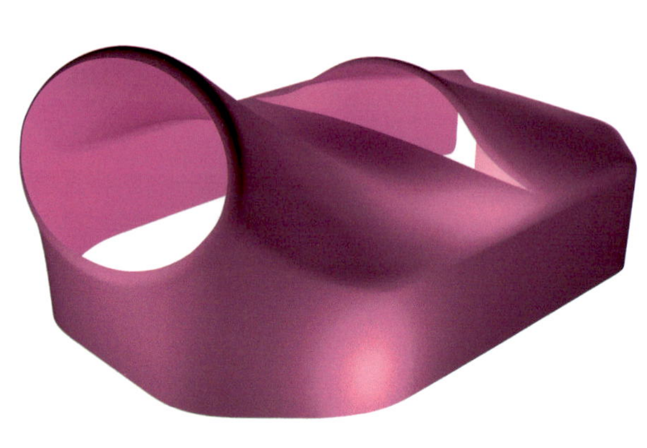
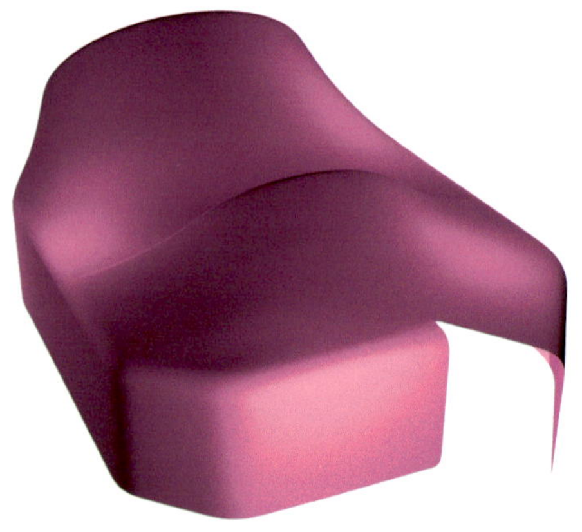

11

12
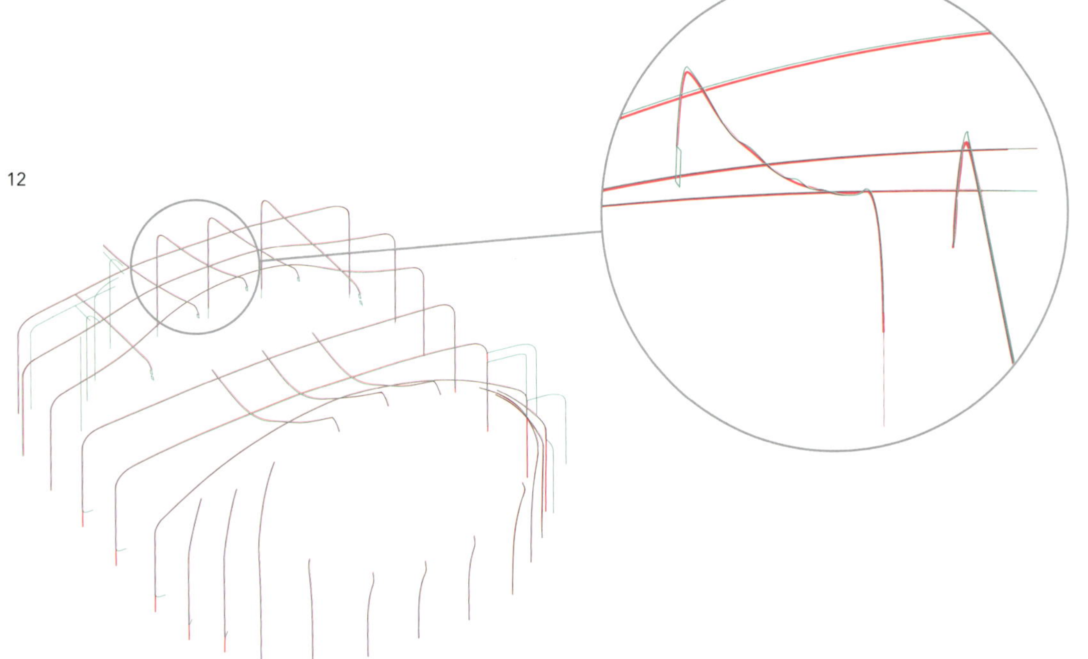

be tangent along the common edge. If the boundary lines are complex, the surface that is generated may be difficult to control and have undesired ripples. It just may not be possible to build a completely smooth surface in this way.

The computer animation industry has developed algorithms that can generate realistic organic shapes starting from only a few control points of a basic mesh. This allows for simplified control of an animated model, as well as simple definition of complex realistic shapes. This smoothing process is based on iterative subdivisions of the initial basic mesh, whereby each new mesh is denser and its vertices tend to belong to a surface that is smooth in all directions.

As can be seen in Figure 10, the final result is a smooth object but is some distance apart from the original geometry because all vertices are shifted to suit the tangency requirements. As the process continues, the variation between successive iterations decreases.

Borrowing from the animation industry, the same idea can be applied to the drawing studio to eliminate the imperfections and difficulties of a manual definition. However, two aspects distinguish this exercise from computer graphics applications: here the initial basic mesh is not given, and the aim is not just to obtain any smooth geometry, but one that is as close as possible to the original manual definition and which essentially only eliminates its imperfections.

By subdividing the initial geometry into a relatively fine mesh corresponding to one of the final iterations, any subsequent iteration would deviate little. With only two iterations, the shape of Figure 11 is created which, to any perceptible degree, is continuous and smooth all around: the differences are small, but at the same time the imperfections have been resolved.

The continuous and smooth surface was important in this case for aesthetic purposes, but its usefulness extends beyond that. A fold would impose a constraint in the way the surface is fabricated as it would require the metal sheets to be spliced; on the other hand, a smooth curvature allows more flexibility in the subdivision of the skin into its basic fabrication parts. A gradual 'smoothed' change of curvature, without sudden variations, also facilitates the forming of the sheets.

STRUCTURAL WEIGHT/COST OPTIMISATION

When provided with stiffeners, the continuous metal envelope also acquires the capacity to function as a primary structure. From a structural design point of view, the ribs should be sized and spaced to provide both the strength to support all design loads, and the stiffness to control live load and thermal deformations, and to prevent the skin from buckling under service conditions.

An optimal spacing can be considered one that is small enough to perform all the above functions and wide enough to limit self-weight, use of material and fabrication works. The optimal spacing would also depend on the structural depth on the stiffeners' cross

10 Smooth reverse of the Loop and Catmull–Clark subdivision.
Demonstration of control of an animated model.

11 CRAB with AKT II, drawing studio, Arts University Bournemouth, Poole, Dorset, UK, 2015.
Subdivision process used for the 3D model definition and resulting surface.

12 CRAB with AKT II, drawing studio, Arts University Bournemouth, Poole, Dorset, UK, 2015.
Comparison of the 3D models before and after smoothing. Compare initial and final surface cuts.

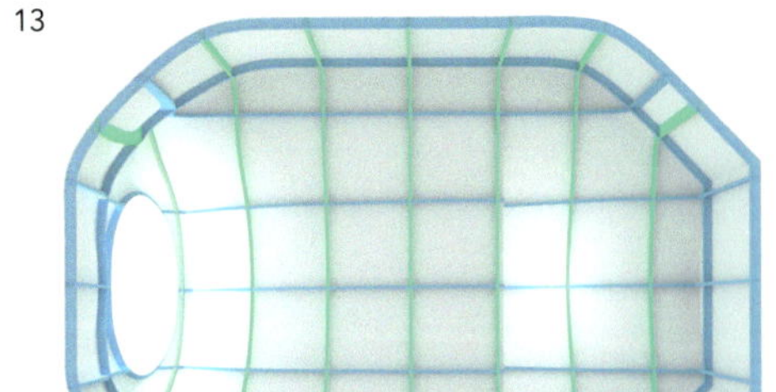

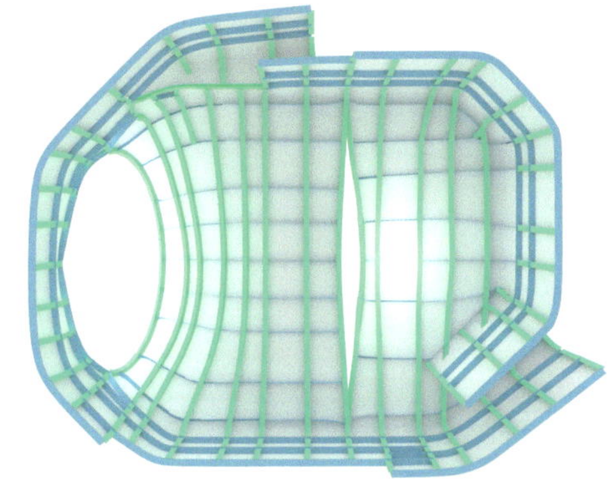

13 CRAB with AKT II, drawing studio, Arts University Bournemouth, Poole, Dorset, UK, 2015.
Alternative stiffener patterns.

14–16 CRAB with AKT II, drawing studio, Arts University Bournemouth, Poole, Dorset, UK, 2015.
Fabrication and pre-assembly by CIG.

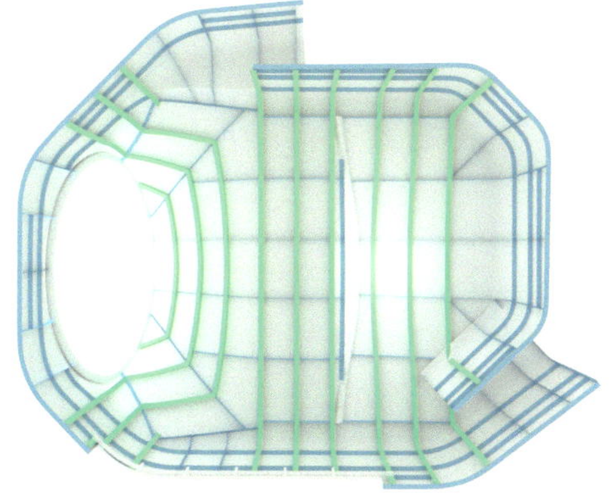

section (for example, the presence of a flange or not) and on the chosen thickness of metal sheets. Of the three different spacings analysed in Figure 12 for the drawing studio, the 1,200 mm one is the lightest that complies with all the performance criteria.

STIFFENERS PATTERN

The pattern of stiffeners is able to influence the loading transfer mechanism, but it is also strongly influenced by the geometry itself. A simple method to generate a stiffening pattern is that of slicing the surface along a defined grid, for example a Cartesian or a polar grid. This method adapts well to regular shapes, as is the case for an aircraft fuselage, usually cylindrical: when arranged along a polar grid, longitudinal ribs are obtained that are always orthogonal to the skin. Transverse circumferential ribs can be generated from cross-planes and are likewise orthogonal to the skin.

With the drawing studio geometry, this ideal scenario cannot be achieved by this method, and severely acute angles between skin and ribs are almost inevitable: these are undesirable because they make welding more difficult, make the structural zone uneven, and complicate the fixing of internal fittings such as insulation and ceilings.

An alternative method, illustrated in Figure 13, generates ribs from cutting planes as in the first method, but the planes are arranged to be approximately perpendicular to the skin, at least along a limited length; when the angle between rib and skin overcomes a set tolerance from being orthogonal, a different plane is used for the adjoining rib.

FABRICATION

The final result depends, to a great extent, on the experience and knowledge of the behaviour of metal when shaped and welded together. Each step exerts a great influence not only on the final product, but also on the way the process is subdivided and

16

17 CRAB with AKT II, drawing studio, Arts
University Bournemouth, Poole, Dorset, UK, 2015.
Site assembly by CIG.

sequenced, because each step introduces some distortions which need to be predicted and accounted for if the final result is to be continuous and smooth.

A complex pre-processing phase of the curvatures in the geometry is used to decide how and where to subdivide it into patches that can be cut out of single metal sheets. The flat shape corresponding to each patch – which is going to be cut out of a steel sheet and then curved – is determined from the final curved geometry by modelling the inverse process and accounting for the deformations that occur during the cold-forming phase. The subdivision into transportable units needs to be established beforehand because it affects the subdivision in patches. A detailed 3D model of each unit is prepared by the fabricator showing the patches, thicknesses, any additional stiffener required for fabrication, and welding preparations; it is then reviewed by the design team.

The stiffeners are more directly obtained via laser-cutting since they have been modelled as flat. Their flange, where present, is usually orthogonal to the web and can therefore be formed by single-curvature of a steel strip, although this curvature may vary along the length. By assembling all the stiffeners together and propping them up, a network is obtained which can be used as the base support and reference for the skin panels. The skin panels can then be positioned and welded on top of the stiffeners.

The patches belonging to each unit are all welded together and to the respective stiffeners, making sure that each new patch is joined to the previous ones tangentially along its borders and without steps. To achieve this, the new patch may need to be slightly forced into position locally to make up for the unavoidable cumulative imperfections of all fabrication phases, before being welded and locked into its unit. Welding itself causes further distortions which are controlled by carefully defining the welding sequencing, as well as adding smaller ribs where required. The tight curvatures around the openings are resolved by welding a tube along the edge, rather than forming steel sheets.

SITE ASSEMBLY

Once welded, the units behave as a single object, but they still influence each other when connected, so that a unit in isolation will have a slightly different shape from the same unit connected to its neighbours. This aspect highlights the importance of keeping at least a minimum level of connection between the units while the building is fabricated, which essentially resulted in a complete factory preassembly of the entire building (Figure 15).

In preparation for the works on site the units were separated, taking their natural shape. The concrete raft and beam foundations were accurately prepared on site by Morgan Sindall, also the main contractor for the project, with pre-set connections positioned to match those of the steelwork baseplates. All units are hoisted into their final position and propped before

18 CRAB with AKT II, drawing studio, Arts
University Bournemouth, Poole, Dorset, UK,
2015.
Building steelwork on site, assembled and fully
propped up by CIG.

any site welding is started. The skin, being an integral part of the structure, allowed the building to be designed as a unified shell; however, at the same time this means that during the assembly all parts need to remain propped up until they are all completely welded together. Once again, the units are likely to require some imposed distortion before being connected to the neighbouring ones in order to recreate the same conditions of the pre-assembled phase, so ensuring that the final result is also a smooth surface along the unit borders.

When the props are removed, the structure acts as a whole, with skin and stiffeners fully working together.

This chapter has tracked the evolution of structures from frames of beams and columns to structural skins, where the cladding contributes to the structural performance of the building, while also functioning as the environmental enclosure and the finished architectural form. This is a trend that poses a number of technological challenges, as the integration of different, sometimes conflicting functions within the same building component stretches the boundaries of what is achievable. Functional integration has an impact on the procurement and maintenance of buildings; the responsibility of the design of a building's components is sometimes simpler to attribute, using a traditional route. However, this also reduces the complexity of interfaces, as well as risk and cost of their coordination on site, and justifies investment in a more integrated design.

TEXT
© 2016 John Wiley & Sons Ltd

IMAGES
Figure 1 © Popperfoto / Getty Images; figure 2 © Bettmann/CORBIS; figure 3 © Luke Hayes; figure 4 © Alex Bilton; figure 5 © Valerie Bennet/AKT II; figures 6, 7, 9, 11, 12, 13 and 14 © AKT II; figure 8 © Valerie Bennet/ AKT II; figure 10 © Reprinted from Javad Sadeghi and Faramarz F Samavati, Smooth Reverse Loop and Catmull-Clark Subdivision, Graphic Models, Vol 73, No 5, September 2011, pp 20_217, with permission from Elsevier; figures 15, 16 and 18 © Cook Robotham Architectural Bureau; figure 17 © CIG Architecture

13 HYBRID SHELLS

JEROEN JANSSEN AND
RICHARD PARKER

For centuries, vaults and arching structures have provided an efficient and economical way of achieving long spans. From the Romans, who introduced arching structures in brick and stone, to the Gothic extremes of slenderness and intricacy, and the Renaissance, where rules and proportion allowed the widespread use of this typology, to the Baroque, where these rules were, again, bent and pushed to the limit. With the advent of the Industrial Revolution and the introduction of steel, trusses provided an alternative to long-span structures, which relied on bending and did not require compressive arched forms. Meanwhile, with the development of reinforced concrete, shell structures, which transfer load through direct compression, found new slender expressions in architecture, replacing traditional masonry.

1

A fundamental shortcoming of thin shells is their limited capacity to resist lateral loads, which becomes increasingly more pronounced as a shell is optimised and reduced in thickness. Through the work of Antoni Gaudí, Pier Luigi Nervi, Felix Candela and Hans Isler, concrete shell structures have reached the peak of slenderness and purity of form. Even with more recent work on digitally form-found organic shells, it seems that a limit has been reached on shell optimisation and, more fundamentally, that questions are raised as to their flexibility in the creation of architectural space and new expressions. The form of shells is strongly constrained by their need to follow some sort of inverted catenary profile in order to achieve a state of pure compression. Based on physical hanging chain models (like those used by Gaudí over a century ago) or their digital equivalent today, the shape of compression shells poses considerable limits on their appearance and ability to respond to complex internal programmatic arrangements.

These classic examples are purist forms of structural minimalism, designed to resist gravity loads in simple compression, keeping depth and volume of material used to the bare minimum. However, they have proven rather inflexible, both in structural and architectural terms. Structurally, their ability to resist asymmetric or lateral loads, such as seismic actions, is reduced the more they are optimised under vertical forces. In architectural terms, their form is limited to a reduced palette of geometries, based on the inverted catenary as the starting point. Influenced by classic catenary forms, ribbing patterns are also more predictable, limiting the range of architectural expressions. Hybrid shells are the result of a multi-parametric design process that contaminates the optimal structural form with spatial, programmatic and aesthetic drivers, which also present the opportunity of optimising material to complex loading combinations of vertical and horizontal dynamic actions.

CONCRETE HYBRID SHELLS

In the past few years, research has been carried out on a large complex of enclosed buildings and open canopies, which extends the notion of pure compressive arching structures to more complex hybrid shells. These hybrid shells are not the traditional, idealistic, engineered expression of the optimal compressive structure, nor are they forms purely shaped by the crafted lines of the architect; these are complex organisations, the integrated response to the multiple requirements of different disciplines. In hybrid shells, neither architecture nor engineering overrides the other. From the outset, many factors influenced the design of the project, such as unforgiving seismic loads that could not be resisted by a laterally weak form, an architect with a strong vision of fine-edged shells, and a series of spatial constraints dictated by a complex set of programmatic requirements. All of these factors pushed and pulled the form of the shells and the project as a whole, until a 'design equilibrium' was reached, satisfying all constraints.

The project essentially consisted of a two-stage design process in which layers of information and geometric rules were built up in response to the various requirements. The first stage consisted

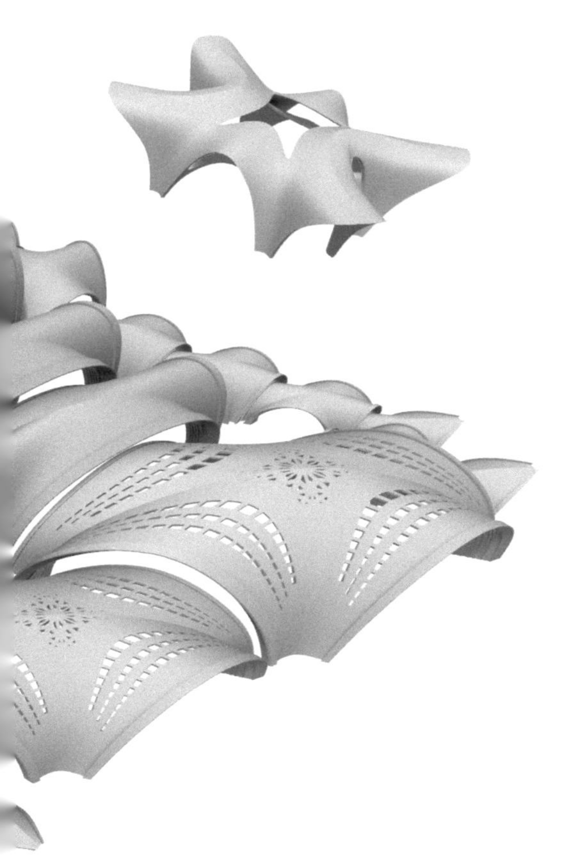

1 Zaha Hadid Architects with AKT II, hybrid shells.
Overview of the whole project, comprising 53 concrete shells of varying size and topology.

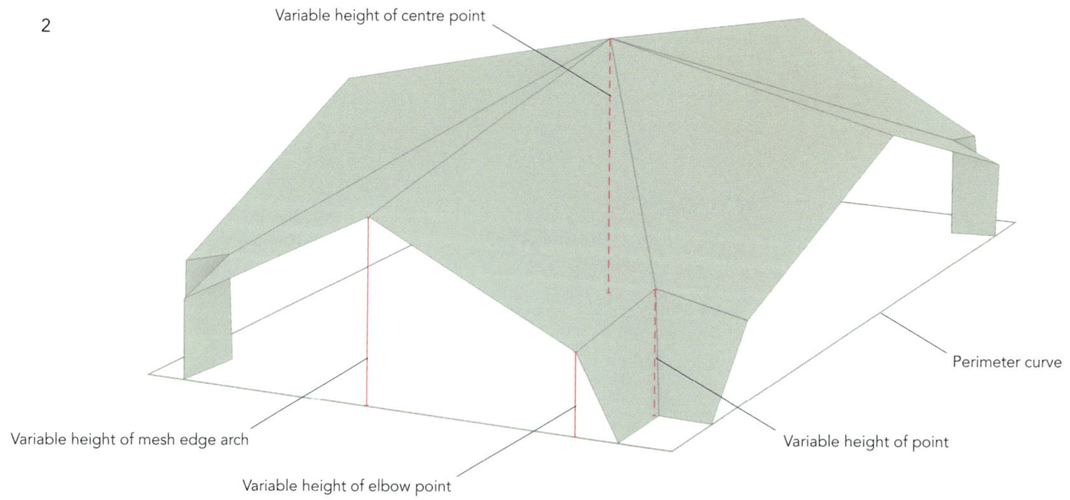

2
Variable height of centre point
Perimeter curve
Variable height of mesh edge arch
Variable height of point
Variable height of elbow point

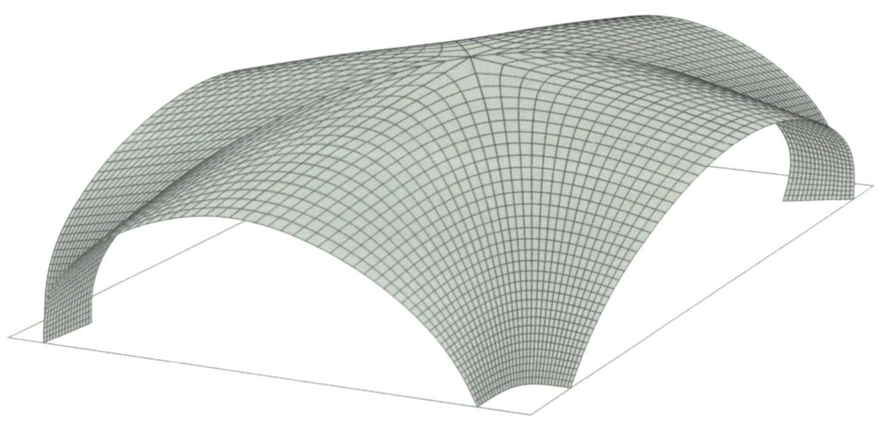

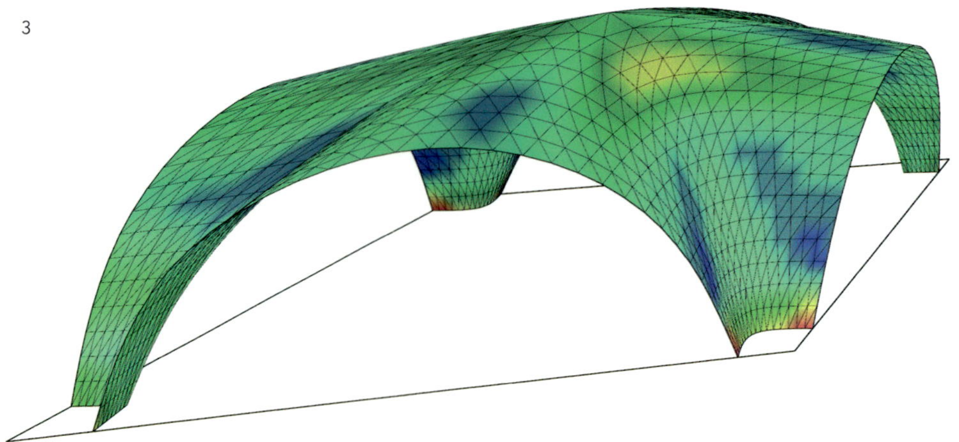

3

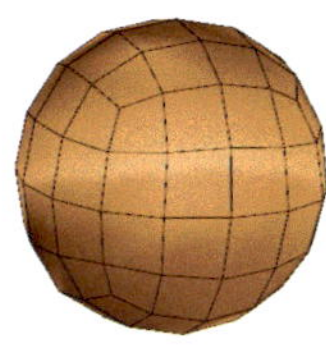
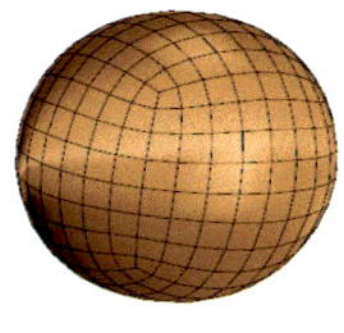

4

2 Zaha Hadid Architects with AKT II, hybrid shells, crude mesh and smooth mesh.
The 'crude mesh' model is the simplest possible quadratic mesh, describing the topology, nodes, creases and seam patterns of the shell geometry. After a process of Catmull–Clark smoothing and subdivision, the resulting shape is a smooth mesh that defines the architectural and structural definition of the shell geometry.

3 Zaha Hadid Architects with AKT II, hybrid shells, geometric algorithm form-finding process.

4 AKT II, Catmull–Clark smoothing and subdivision algorithm.
Diagram showing the process, going from no subdivisions on the left, to three levels of subdivision on the far right.

of the form- and pattern-finding of the shells, determining their primary setting out; the second was the development of a detailed, coordinated model of the project, incorporating the requirement of each discipline. To achieve equilibrium in this multi-parametric design, a common platform was developed that allowed the sharing of data across the project.

The process started with the definition of the shells' topology and proceeded with their form-finding, ensuring that efficiency could be achieved within forms of predetermined geometric characteristics. A 'crude mesh', the simplest possible quadratic mesh, describing the topology of the shells – the nodes, creases and seam patterns – was used as the controlling object. Through a process of Catmull–Clark smooth subdivision, the crude mesh was processed into a smooth mesh, which was used to define the precise form of the shells, their curvature and dimensions. With a detailed understanding of the mathematical principles behind this subdivision algorithm, design engineers were able to manipulate the smooth-subdivision process, introducing the ability to create custom boundary conditions. For example, the positions of certain edge vertices in the crude mesh were constrained along planes, ensuring that the corresponding vertices of the smooth mesh lay along the same boundaries.

Once the principles underlying the relationship between the final smooth form and its originating crude mesh were established, the process of analysis and optimisation could take place. A multi-parametric form-finding process for the shells was set up, aimed at: satisfying geometric requirements relating to the constructability of the concrete surfaces; catering for multiple structural load conditions; allowing for spatial constraints deriving from programmatic arrangements; and allowing control of the appearance of the architectural form. A parametric 'definition', working in series with finite element solvers, was the key to allowing this integrated process. Through the use of such tools, which embedded engineering intelligence as well as manual control within the same form-finding process, design engineers were able to assess the optimal form of hundreds of shell iterations. This was based on a genetic algorithm, designed to arrive at a suitable solution with the minimum number of iterations of a pre-set range of geometric parameters.

While a logic of structural efficiency was engrained within the process of form-finding, the designer maintained control by

5 Zaha Hadid Architects with AKT II,
hybrid shells.
Overview of the evolution of the shape
of the smooth parent shell during the
design process.

6 Zaha Hadid Architects with AKT II,
hybrid shells.
Following the offset zones, the edges
of the surfaces are restrained on
3D planes, generating curves at the
intersection with the surfaces. These
curves will define the edge definitions
of the shell geometry.

5

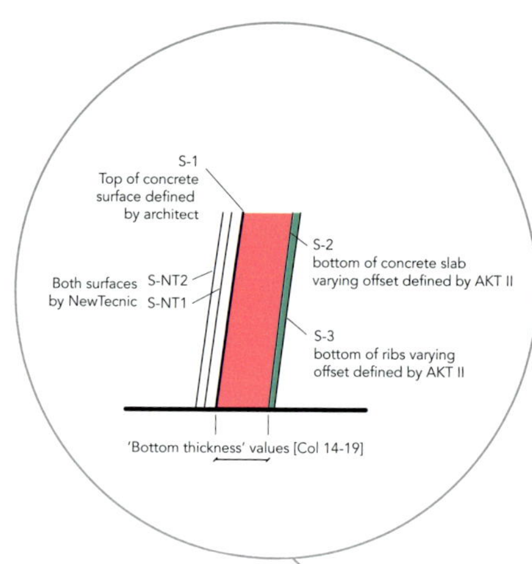

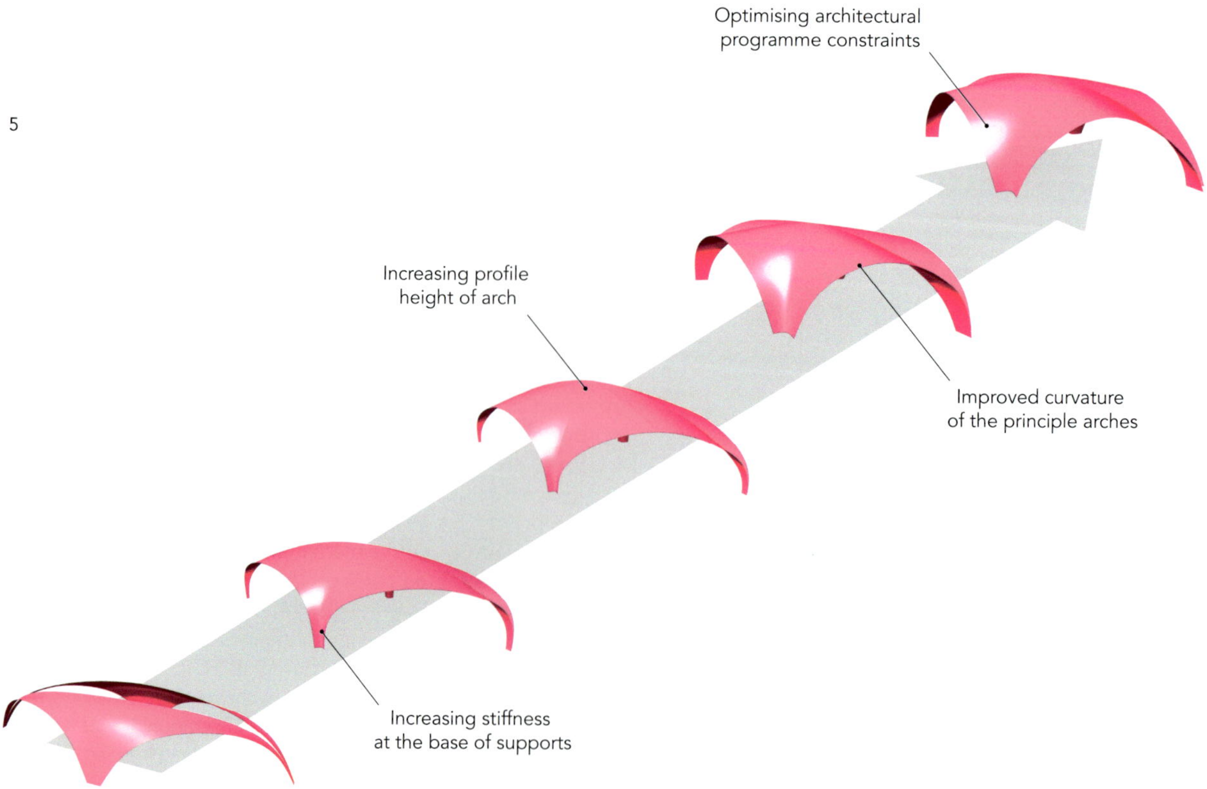

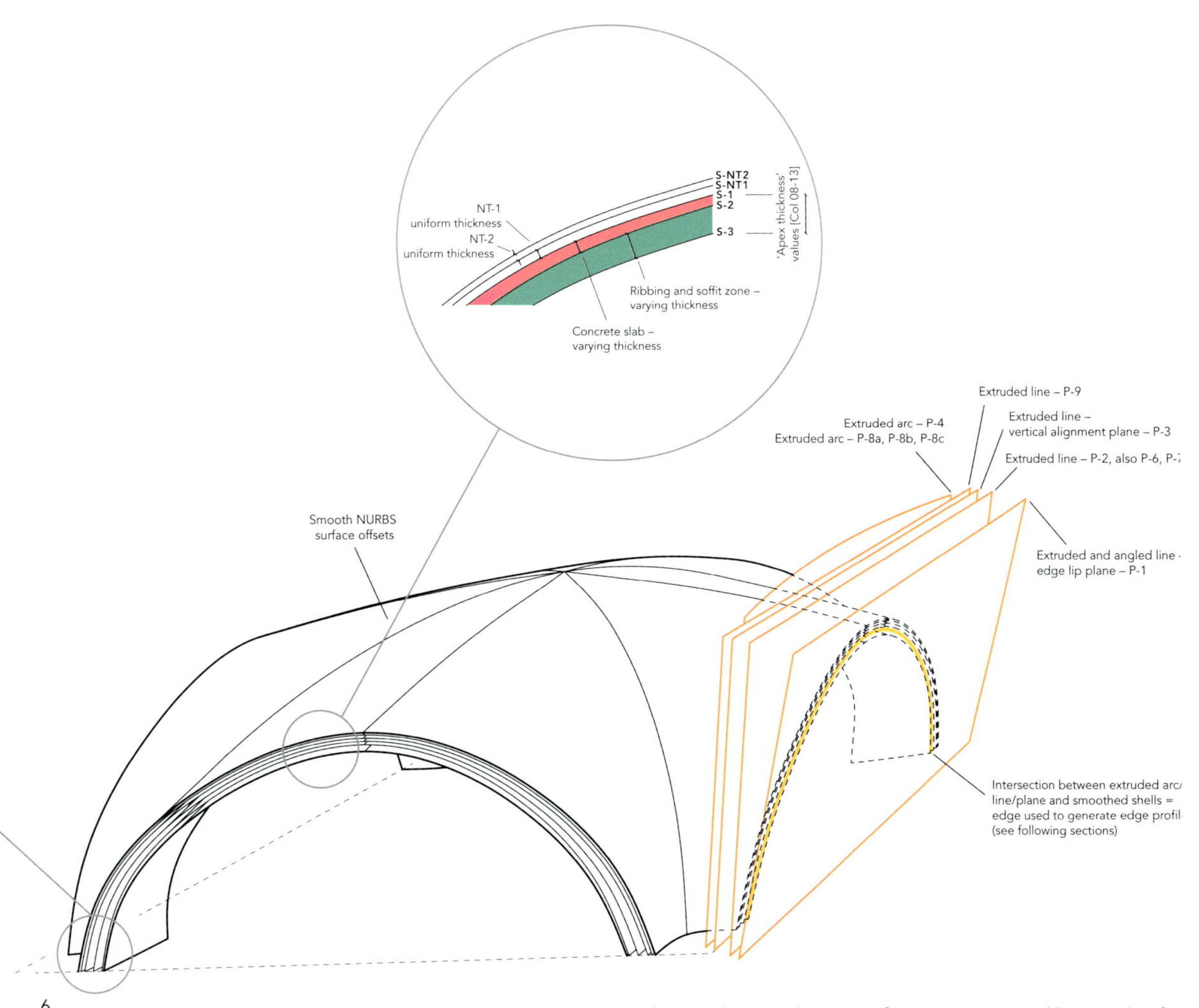

manipulating the initial range of parameters, and hence the forms that could be achieved. By integrating analytical form-finding with a user-driven crafted tool, an efficient convergence between architectural design and engineering efficiency was achieved. Rigorous analytical optimisation procedures were combined with intuitive crafted modelling techniques within an analogue-to-digital design feedback-loop. This method is more than the merging of a traditional purist engineered approach to form-finding with a post-rationalisation of the irrational preconceived architectural form; it is a sophisticated, integrated design process of multiple parameters, from aesthetic and programmatic ones to those associated with structural efficiency or constructability.

The outcome of the form-finding process was an optimised form, defined by a single smooth setting-out mesh. From this, a polysurface of quadratic NURBS with continuous and smooth isoparametric curves could be derived, providing the geometric framework for the mapping of ribbing, fenestration patterns and other features onto the shells. Structural ribbing patterns were applied to the 3D surfaces with varying depths and widths, based

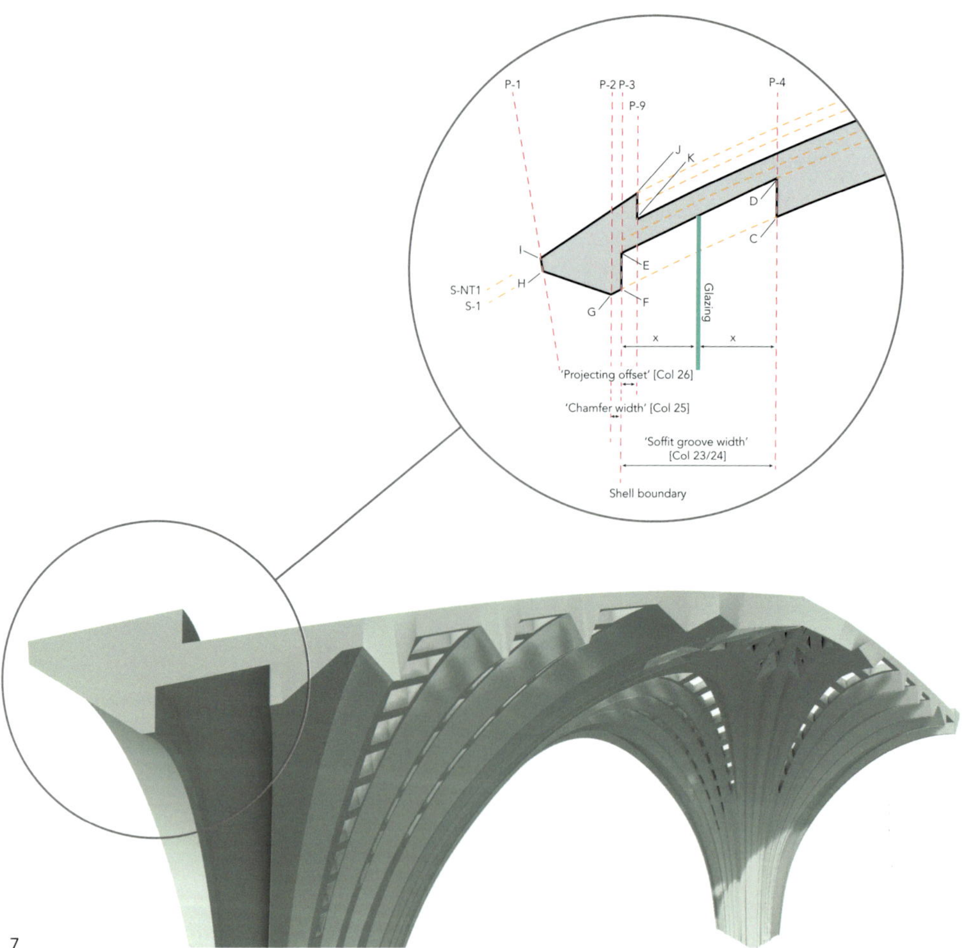

7

8

Thicknesses — Measured From S-1 (top setting out surface) [mm]

1	2 Split Type [long axis or diag]	3 Shell ID	4 Description	5 Type ID	6 Type of Shell/Ribs	7 Mirror Shell? [Yes or No]	8 APEX S-2	9 APEX S-3	10 APEX S-F	11 APEX S-NT1	12 APEX S-NT2	13 APEX S-A	14 BOTTOM S-2	15 BOTTOM S-3	16 BOTTOM S-F	17 BOTTOM S-NT1	18 BOTTOM S-NT2	19 BOTTOM S-A
1	A	S0a	Reception Sister	1A	Ribs with slab openings	Y	200	800	0	-210	-360	100	900	1000	0	-210	-360	450
2	A	S1a	Banquet Sister	1A	Ribs with slab openings	Y	200	800	0	-210	-360	100	900	1000	0	-210	-360	450
3	A	S2a	Presidential Offices	1B	Ribs	Y	200	800	0	-210	-360	100	900	1000	0	-210	-360	450
4	A	S3a	Presidential Offices	1B	Ribs	Y	200	800	0	-210	-360	100	900	1000	0	-210	-360	450
5	A	S7a	Circulation	4	Smooth Underside	Y	250	250	0	-210	-360	125	500	500	0	-210	-360	250
6	A	S8a	Circulation	4	Smooth Underside	Y	250	250	0	-210	-360	125	500	500	0	-210	-360	250
7	A	S9a	Circulation	4	Smooth Underside	Y	250	250	0	-210	-360	125	500	500	0	-210	-360	250
8	A	C1a	Esplanade Shells	4	Smooth Underside	Y	250	250	-35	-210	-260	125	600	600	60	-210	-260	300
9	A	C2a	Esplanade Shells	4	Smooth Underside	Y	250	250	-35	-210	-260	125	600	600	75	-210	-260	300
10	A	C3a	Esplanade Shells	4	Smooth Underside	Y	250	250	-35	-210	-260	125	600	600	90	-210	-260	300
11	A	C4a	Esplanade Shells	4	Smooth Underside	Y	250	250	-10	-210	-260	125	600	600	75	-210	-260	300
12	A	C5a	Esplanade Shells	4	Smooth Underside	Y	250	250	-30	-210	-260	125	600	600	80	-210	-260	300
13	A	C6a	Esplanade Shells	4	Smooth Underside	N	250	250	-35	-210	-260	125	600	600	80	-210	-260	300
14	A	C7a	Esplanade Shells	4	Smooth Underside	N	250	250	-35	-210	-260	125	600	600	85	-210	-260	300
15	A	C8a	Esplanade Shells	4	Smooth Underside	N	250	250	-25	-210	-260	125	600	600	70	-210	-260	300
16	A	C9a	Esplanade Shells	4	Smooth Underside	N	250	250	-20	-210	-260	125	600	600	70	-210	-260	300
17	A	C12a	Esplanade Shells	4	Smooth Underside	Y	250	250	-25	-210	-260	125	600	600	65	-210	-260	300
18	A	C13a	Esplanade Shells	4	Smooth Underside	Y	250	250	-25	-210	-260	125	600	600	70	-210	-260	300
19	A	C32a	Esplanade Shells	4	Smooth Underside	Y	250	250	-35	-210	-260	125	600	600	75	-210	-260	300
20	A	C33a	Esplanade Shells	4	Smooth Underside	Y	250	250	-35	-210	-260	125	600	600	85	-210	-260	300
21	A	C35a	Esplanade Shells	4	Smooth Underside	Y	250	250	-35	-210	-260	125	600	600	60	-210	-260	300
22	B	C30	Circulation	4	Smooth Underside	N	400	400	0	-210	-360	200	600	600	0	-210	-360	300
23	B	S40	Presidential Offices	3B	Diagonal Grid w. Round Smooth Soffits	N	500	800	0	-210	-360	250	750	800	0	-210	-360	375
24	B	S50	Official Entrance	2	Diagonal Grid w. Shattered Diamond Vaults	N	300	500	0	-210	-360	150	450	500	0	-210	-360	225
25	B	S60	Official Entrance	2	Diagonal Grid w. Shattered Diamond Vaults	N	300	500	0	-210	-360	150	450	500	0	-210	-360	225
26	B	C70	Circulation	4	Smooth Underside	N	400	400	0	-210	-360	200	600	600	0	-210	-360	300
27	C	T0a	Esplanade Shells (Cantilever)	4	Smooth Underside	Y	250	250	5	-210	-260	125	600	600	50	-210	-260	300
28	C	T6b	Esplanade Shells (Cantilever)	4	Smooth Underside	N	250	250	5	-210	-260	125	600	600	65	-210	-260	300
29	C	T11a	Esplanade Shells (Cantilever)	4	Smooth Underside	Y	250	250	0	-210	-260	125	600	600	65	-210	-260	300
30	C	T14a	Esplanade Shells (Cantilever)	4	Smooth Underside	N	250	250	0	-210	-260	125	600	600	60	-210	-260	300
31	C	T10a	Esplanade Shells (Cantilever)	4	Smooth Underside	N	250	250	-30	-210	-260	125	600	600	160	-210	-260	300
32	C	T31a	Esplanade Shells (Cantilever)	4	Smooth Underside	Y	250	250	0	-210	-260	125	600	600	65	-210	-260	300
33	C	T34a	Esplanade Shells (Cantilever)	4	Smooth Underside	Y	250	250	-35	-210	-260	125	600	600	180	-210	-260	300
34	C	T36a	Esplanade Shells (Cantilever)	4	Smooth Underside	Y	250	250	-5	-210	-260	125	600	600	65	-210	-260	300
35	A	S0b	Reception Hall	1A	Ribs with slab openings	N												
36	A	S1b	Banquet Hall	1A	Ribs with slab openings	N												
37	A	S2b	Presidential Offices	1B	Ribs	N												
38	A	S3b	Presidential Offices	1B	Ribs	N												
39	A	S7b	Circulation	4	Smooth Underside	N												
40	A	S8b	Circulation	4	Smooth Underside	N												
41	A	S9b	Circulation	4	Smooth Underside	N												
42	A	C1b	Esplanade Shells	4	Smooth Underside	N												
43	A	C2b	Esplanade Shells	4	Smooth Underside	N												
44	A	C3b	Esplanade Shells	4	Smooth Underside	N												
45	A	C4b	Esplanade Shells	4	Smooth Underside	N												
46	A	C5b	Esplanade Shells	4	Smooth Underside	N												
47	A	C12b	Esplanade Shells	4	Smooth Underside	N												
48	A	C13b	Esplanade Shells	4	Smooth Underside	N												
49	A	C32b	Esplanade Shells	4	Smooth Underside	N												
50	A	C33b	Esplanade Shells	4	Smooth Underside	N												
51	A	C35b	Esplanade Shells	4	Smooth Underside	N												
52	C	T0b	Esplanade Shells (Cantilever)	4	Smooth Underside	N												
53	C	T11b	Esplanade Shells (Cantilever)	4	Smooth Underside	N												
54	C	T31b	Esplanade Shells (Cantilever)	4	Smooth Underside	N												
55	C	T34b	Esplanade Shells (Cantilever)	4	Smooth Underside	N												

Edge Lip, Edge Profile, Glazing and soffit data (BLACK = ZHA VALUE / Glazing inside Shell; GREEN = AKTII value to ensure Shell overlap; BLUE = ZHA VALUE; RED = NT VALUE)

Shell ID	20 Edge Lip APEX	21 Edge Lip BOTTOM	22 Edge Lip Type [clockwise from top]	23 Soffit Groove Min Width [mm] PURPLE= Controlled BY S-G Vals	24 Soffit Groove Max Width [mm] GREEN= Controlled BY S-G Vals (changed)	25 Chamfer Width [mm]	26 Projecting Offset [mm]	30 Glazing Min Width [mm]	31 Glazing Max Width [mm]	33 Long Axis Typology	34 Soffit Type [ribs, facets, smooth]	35 Soffit Type [smooth or ridged]	36 Cantilever Edge	37 Tolerance Type
S0a	1500,1500,1500,1500	600,600,600,600	1,1,1,4	1000,1000,1000,1000	1200,1200,1200,1200	120	-250	750,700,700,500	1200,1200,1200,750	Straight	0	A	N,N,N,N	A
S1a	1500,1500,1500,1500	600,600,600,600	4,2,1,1	1000,1000,1000,1000	1200,1200,1200,1200	120	-250	500,500,650,650	750,750,1200,1200	Straight	0	A	N,N,N,N	A
S2a	1500,1500,1500,1500	600,1200,600,600	4,_,2,1	1000,1000,500,1100	500,1200,600,1400	120	-250	500,1200,500,1100	750,1400,750,1400	Arc	1	A	N,N,N,N	A
S3a	1500,1500,1500,1500	600,600,600,600	4,1,2,1	1000,1200,500,1100	1200,1400,1200,1400	120	-250	500,1150,500,1100	750,1400,750,1400	Arc	1	A	N,N,N,N	A
S7a	1000,1500,1500,1000	500,400,500,400	4,1,2,1	500,800,500,800	1200,1600,1200,1400	120	-250	500,650,500,650	750,1000,750,1000	Straight	3	A	N,N,N,N	B
S8a	1000,1000,1000,1000	500,400,500,400	4,1,2,1	800,800,750,800	1000,1000,750,1000	120	-250	500,650,500,500	750,1000,750,750	Straight	3	A	N,N,N,N	B
S9a	1000,1000,1000,1000	400,500,400,400	1,4,_,2	800,800,750,500	1000,1000,750,750	120	-250	650,500,500,500	1100,750,750,750	Straight	3	B	N,N,N,N	B
C1a	500,500,500,500	80,80,80,80	5,5,5,5				-200			Straight	3	A	N,N,N,N	A
C2a	500,500,500,500	80,80,80,80	5,5,5,5				-200			Straight	3	A	N,N,N,N	A
C3a	500,500,500,500	80,80,80,80	5,5,5,5				-200			Straight	3	A	N,N,N,N	A
C4a	500,500,500,500	80,80,80,80	5,5,5,5				-200			Straight	3	A	N,N,N,N	A
C5a	500,500,500,500	80,80,80,80	5,5,5,5				-200			Straight	3	A	N,N,N,N	A
C6a	500,500,500,500	80,80,80,80	5,5,5,5				-200			Straight	3	A	N,N,N,N	A
C7a	500,500,500,500	80,80,80,80	5,5,5,5				-200			Straight	3	A	N,N,N,N	A
C8a	500,500,500,500	80,80,80,80	5,5,5,5				-200			Straight	3	A	N,N,N,N	A
C9a	500,500,500,500	80,80,80,80	5,5,5,5				-200			Straight	3	A	N,N,N,N	A
C12a	500,500,500,500	80,80,80,80	5,5,5,5				-200			Straight	3	A	N,N,N,N	A
C13a	500,500,500,500	80,80,80,80	5,5,5,5				-200			Straight	3	A	N,N,N,N	A
C32a	500,500,500,500	80,80,80,80	5,5,5,5				-200			Straight	3	A	N,N,N,N	A
C33a	500,500,500,500	80,80,80,80	5,5,5,5				-200			Straight	3	A	N,N,N,N	A
C35a	500,500,500,500	80,80,80,80	5,5,5,5				-200			Straight	3	A	N,N,N,N	A
C30	500,500,1000,500	80,80,80,80	5,5,5,5				-250			Smooth	3	B	N,N,N,N	A
S40	1500,1500,1500,1500	600,600,400,400	1,3,2,3	1100,1000,1000,1000	1400,1200,1200,1200	120	-250	1100,500,1200,500	1400,750,1550,750	Smooth	4	B	N,N,N,N	A
S50	1000,1000,1000,1000	400,400,500,400	_,1,4,1	800,800,250,800	900,1000,500,1000	120	-250	500,700,500,700	750,1000,750,1000	Smooth	2	B	N,N,N,N	B
S60	1000,500,1700,1000	500,500,500,500	2,2,1,2	500,500,700,500	800,800,1000,800	120	-250	500,500,700,500	750,750,1000,750	Smooth	2	B	N,N,N,N	B
C70	1000,500,500,500	80,80,80,80	5,5,5,5				-200			Smooth	3	B	N,N,N,N	A
T0a	500,500,500,500	80,80,80,80	5,5,5,5				-200			Straight	3	A	N,N,Y,Y	A
T6b	500,500,500,500	80,80,80,80	5,5,5,5				-200			Straight	3	A	Y,Y,N,N	A
T11a	500,500,500,500	80,80,80,80	5,5,5,5				-200			Straight	3	A	N,N,Y,Y	A
T14a	500,500,500,500	80,80,80,80	5,5,5,5				-200			Straight	3	A	N,N,Y,Y	A
T10a	500,500,500,500	80,80,80,80	5,5,5,5				-200			Straight	3	A	N,N,Y,Y	A
T31a	500,500,500,500	80,80,80,80	5,5,5,5				-200			Straight	3	A	Y,Y,N,N	A
T34a	500,500,500,500	80,80,80,80	5,5,5,5				-200			Straight	3	A	Y,N,N,Y	A
T36a	500,500,500,500	80,80,80,80	5,5,5,5				-200			Straight	3	A	Y,N,N,Y	A

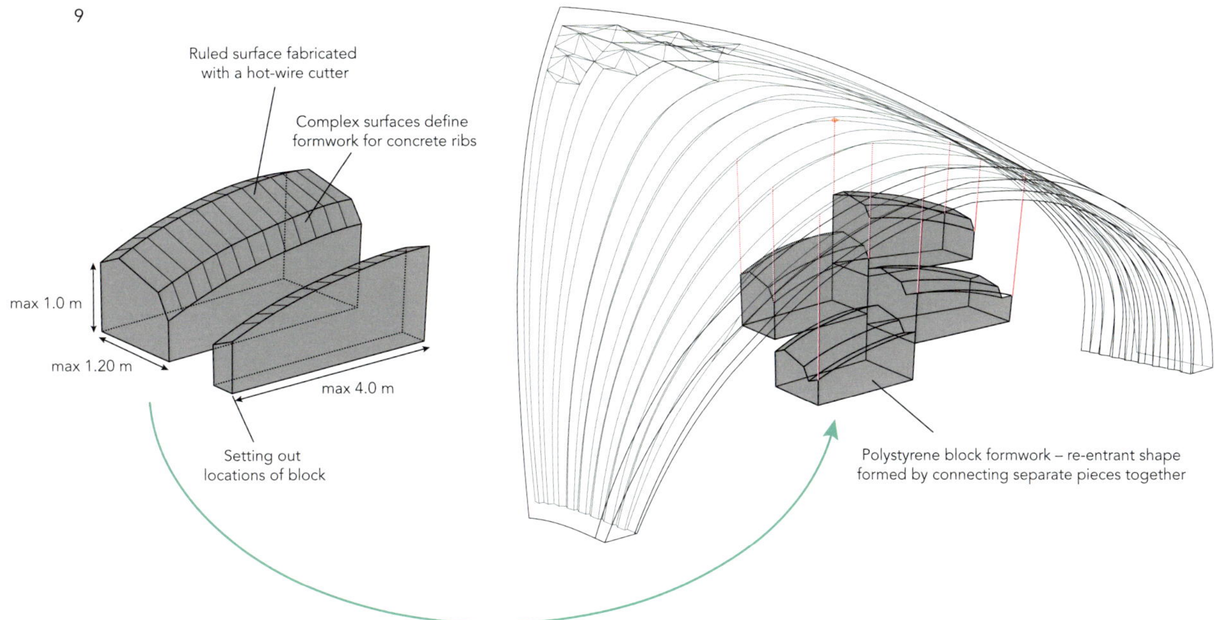

7 Zaha Hadid Architects with AKT II, hybrid shells.
Mapping of the ribbing patterns onto smooth setting-out polysurface.

8 Zaha Hadid Architects with AKT II, hybrid shells.
Overview of the spreadsheet with all input parameters and values that feed into and steer the shared parametric geometry definition.

9 Zaha Hadid Architects with AKT II, hybrid shells, study on the method of construction for formwork.
The complex concrete surfaces are generated by ruled surfaces which could be fabricated by a hot-wire cutter.

on local stress levels. This resulted in thick solid landing points for channelling large forces to the ground, and gradually thinner and lighter ribs at the crown to reduce mass, weight and seismic forces. This object-oriented mapping of ribs and other complex details proved to be a powerful means of transferring a catalogue of simple details from a 'parent' surface to a series of 'child' surfaces of equivalent topological definition. In a similar way, four families of ribbing patterns could be tested on different shells and matched with the more appropriate geometries.

The ability to communicate beauty and efficiency in this iterative process of design and optimisation – from the outline of the form to the detail of fillets and chamfers of concrete ribs – was essential in order to move towards design decisions and continue the layering of detail into the project. This was particularly true when conflicting parameters had to be weighted and assessed against one another. Geometric rationalisation of the form increased in importance once the practicalities of forming simple shuttering moulds were introduced within the process. To achieve a reasonable construction programme, forms had to be simplified to ruled surfaces buildable with robotically wire-cut polystyrene moulds, rather than milled into more complex forms. This did not meet the structural optimum, but was within an acceptable tolerance range.

With a rigorous underlying data-structure and a powerful visual 3D model encapsulating that logic, the team was able efficiently to communicate the hierarchy and relationship between layers of the project, and create a complex but smooth collaborative workflow. A document was created called the 'geometric statement', which summarised the relationships between parameters, from the input ones to the resulting ones. The project, in its complexity, is

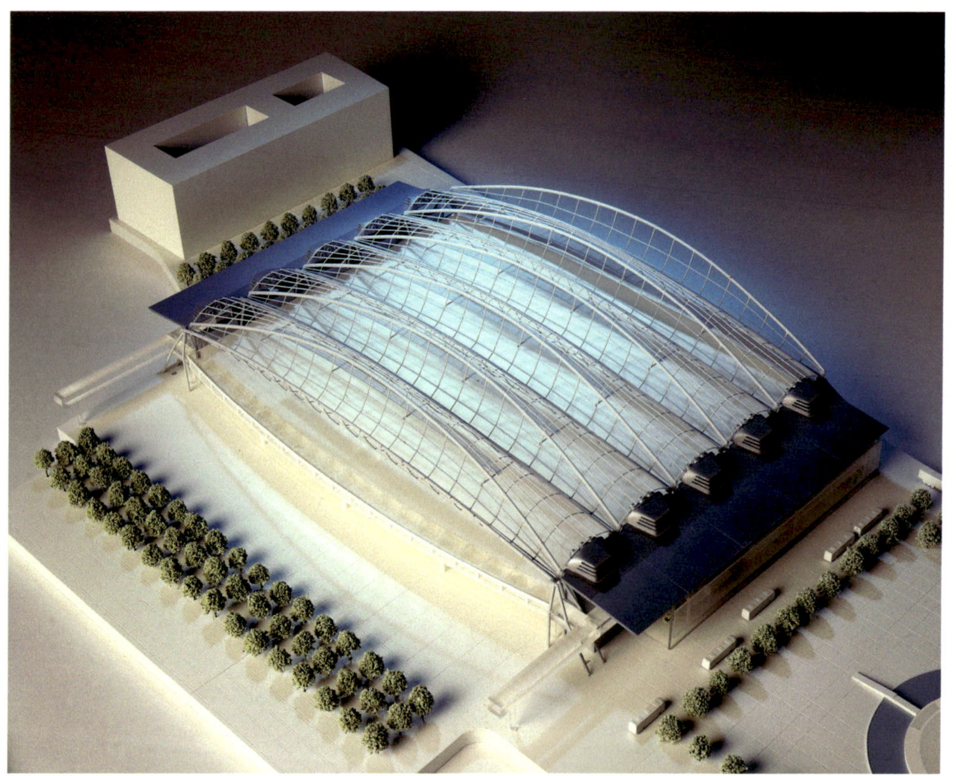

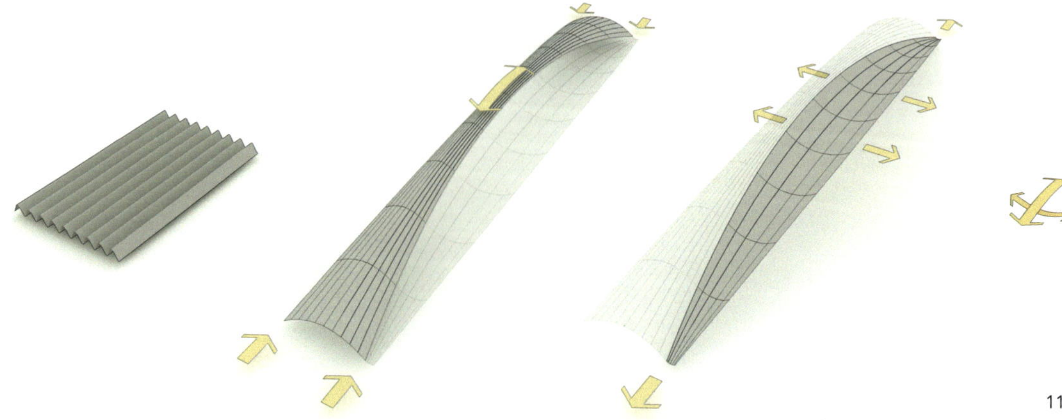

10 Grimshaw, Messehalle 3, Frankfurt,
Germany, 2001.
Model showing the undulating structural
surface.

11 Grimshaw, Messehalle 3, Frankfurt,
Germany, 2001.
Sketches illustrating the undulating structural
surface of the roof with its compression shells
and tensile nets.

the hybrid balance of contrasting forces, values and parameters aimed at producing an integrated, holistic efficient whole. The project redefines the relationship between architect and engineer, maximising their creative and analytical strengths and abilities in the achievement of the collaborative final design. While specific and bespoke, the processes and tools developed are based on elements, objects and components which can be generalised, extended and applied to other projects by a process of assembly and reorganisation.

STEEL HYBRID SHELLS

In the evolution of compression structures – from masonry vaults to reinforced-concrete shells – towards the end of the 19th century gridshells emerged from the use of standard steel profiles or engineered timber beams to form doubly curved lattices. Vladimir Shukhov pioneered this typology in the construction of exhibition pavilions for the All-Russia Industrial and Art Exhibition of 1896 in Nizhny Novgorod. Other classic examples of gridshell structures include Pier Luigi Nervi's 1938 hangar, where a ribbing lattice structurally collaborates with the cladding skin, and Frei Otto's 1975 Mannheim Multihalle, where straight timber joists are pushed into a pure compressive form. These early examples of gridshell structures were based on structural form-finding methods similar to the ones employed for concrete shells; their forms are driven by the stress flows in inverted hanging chain models.

In more recent years, these classic forms have been 'contaminated' by the introduction of other form-finding parameters and materials. In the Pompidou Museum in Metz by Shigeru Ban, for example, the shape of the layered timber lattice is form-found between 'programmatic boxes' that act as rigid boundary conditions, rather than being supported on level ground. In other projects, the architect uses bamboo boards and cardboard tubes interweaving reciprocal arrangements, such as in the Forest Park pavilion in St Louis. In these projects, pure compressive load-paths are combined with tension-filled and flexural behaviour to increase complexity of the form and of visual expression.

12 Grimshaw, Messehalle 3, Frankfurt, Germany, 2001.
Aerial view with its 165 m clear span.

12

In the Frankfurt Messehalle, designed by Grimshaw, an undulating surface is used to achieve a 165 m span over a column-free exhibition hall. The structure, consisting of a single-layer lattice of welded steel tubes, is a hybrid of compression shells and tension-nets within the same smooth surface. Through form alone, the arching system is rendered independent of large ground restraints and foundations by containing within the structure itself the large thrusting forces generated by its shallow arching profile. In other words, the surface acts both as a compression shell and as its tying tensile restraint.

To achieve the balance between tension and compression within a single-layered structure, a complex form-finding process was used based on bespoke software, simulating the dynamic relaxation of the structure under gravity loads. This allowed the prediction of its final relaxed form starting from a pre-set propped position. This method was not only to determine the final geometry of the structure from its un-deformed fabricated shape, but also to eliminate the need for complex jacking processes during its construction on site. Pre-stress in the tension-nets at the valleys was induced by the spread of the compression shells at the ridges, which allowed the structure to slide at the supports under gravity loads. Eventually, the system was 'locked' at the support, ensuring that any varying wind loads would be transferred directly to the A-frames at each end of the roof structure.

CONCLUSION

The work on both the concrete shells and the Frankfurt Messehalle is the result of hybrid form-finding processes that do not account for structural efficiency alone, even in a typology where this is critical. Other aspects are catered for, such as economy of fabrication and practicality of construction. These are all fed back into the design process and, through the use of interactive tools, they allow the development of complex multi-parametric solutions that carry the efficiency of automation as well as the creative act of intuition and imagination.

TEXT

IMAGES

14 TENSILE STRUCTURES

DIEGO CERVERA DE LA ROSA AND ALESSANDRO MARGNELLI; JAMES KINGMAN

LARGE-SCALE TENSILE STRUCTURES

Diego Cervera de la Rosa and Alessandro Margnelli

Tensile structures are the result of the designer's process to find optimally performing minimal surfaces. Their shape is generated by the equilibrium of tension forces and the manner by which these forces flow into the structure. This flow is a function of the surface's curvature, determined by the amount of tensile stress in the system.

The simplest tensile structure consists of a cable. Under the influence of gravity, cables fall into a curve called a 'catenary'. The surface created by the revolution of a catenary curve about a horizontal axis is a 'catenoid'. Catenoids belong to a family of mathematical surfaces known as minimal surfaces, which are characterised by 'zero mean curvature', ie, with the minimal surface between edges.

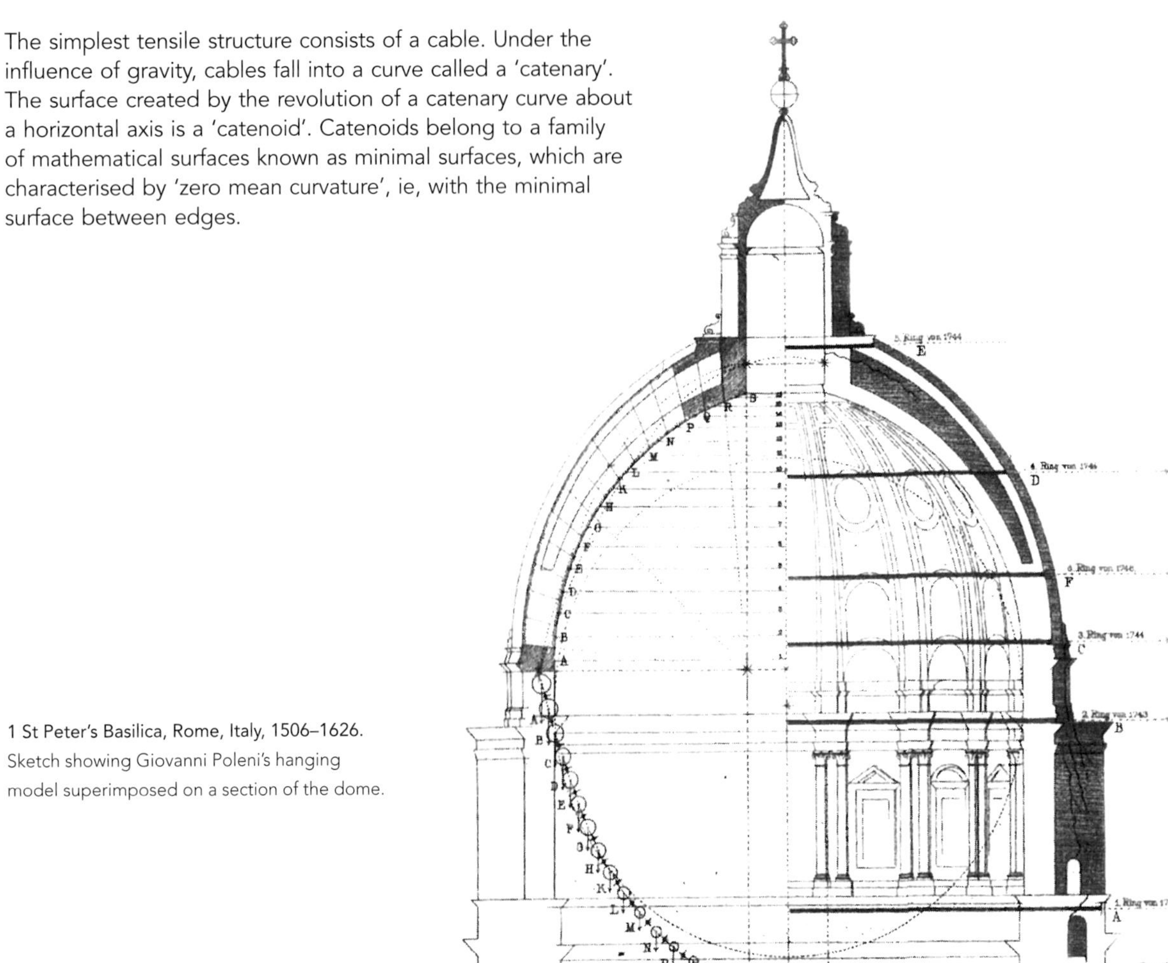

1 St Peter's Basilica, Rome, Italy, 1506–1626. Sketch showing Giovanni Poleni's hanging model superimposed on a section of the dome.

The problem of analysing a catenary is not new, having been studied by several scientists and mathematicians in the past. In 1675, the English natural philosopher, Robert Hooke, published the book *A Description of Helioscopes and Some Other Instruments*,[1] where he inserted several phrases, of which two were later identified as the basics of structural engineering. The first, *'ut tensio, sic vis'*, expresses the proportion between force and extension. The second, *'ut pendet continuum flexile, sic stabit contiguum rigidum inversum'*, states the analogy between an arch and a catenary. In 1691, Gottfried Liebniz, Christiaan Huygens and Johann Bernoulli formulated the same solution and expressed this as a mathematical equation to describe the problem.

Expressed diagrammatically by Giovanni Poleni (1748), the bijective nature between compression and tension (arch and catenary) was widely used to create pure compression arches from models of hanging chains.

Their simplicity has made tensile structures attractive to both engineers and architects. Double curvatures of translational or rotational surfaces (synclastic or anticlastic) and, lately, complex free-forms, have pushed design and analysis towards a common point where architects and engineers have had to work closely together to define the form and pattern of structural elements. The synergy of Ted Happold and Frei Otto, and Cecil Balmond and Alvaro Siza, resulted in projects such as the Multihalle Mannheim (1975) and the Expo '98 Portuguese National Pavilion (Figure 4).

GEOMETRIC PRINCIPLES AND FORMS
Classical forms are typically based on symmetry, and generated through geometric extrusion, translation, and the rotation of curves in space. The form of tensile structures is produced by a process of form-finding that ensures equilibrium of the tensile forces; in these kinds of geometries, the curvature directly affects the behaviour/ stiffness of the structure.

In structures where the curvature defines the geometry, such as arches and catenaries, the curvature directly affects the internal forces and the thrust at the supports: the bigger the curvature, the stiffer the structure. For surfaces, the curvature can't be defined with a single value, so Gaussian curvature is used: this is positive when the U and V curves bend in the same direction (synclastic), and negative when they bend in opposite directions (anticlastic). In the family of anticlastic surfaces, one type has important benefits when used in structural design: the minimal surface. Given a fixed boundary, a minimal surface spans with the minimal energy/minimal area.

A classic example of synclastic surfaces is suspended structures, typically characterised by catenary forms, where the vertical loads are balanced by means of elements working in tension and base their stability under dominant vertical external loads (such as cladding). The nature of the structural behaviour (tension only) allows the use of slender elements to their maximum capacity, as no second order buckling effects take place. This allows the spanning of long distances without considerably increasing the weight. However, suspended structures are extremely sensitive to unbalanced loadings or dynamic effects, since there are no elements restraining the upwards movement. These effects can be avoided by increasing the dead load acting on the structure.

Pre-tensioned anticlastic structures present an evolution of suspended structures. They achieve stiffness independently from external gravity loads by means of internal pre-stress forces, applied in advance of any external loads. This produces structures with an inherent capacity for resisting external actions, allowing the dead weight to be reduced to a bare minimum. Their erection process is not as straightforward as for suspended structures, since the components need to be jacked into a pre-stressed state before the external loads are applied. Jacking is a complex process that needs to be carried out iteratively to achieve the desired stress configuration without affecting the geometry.

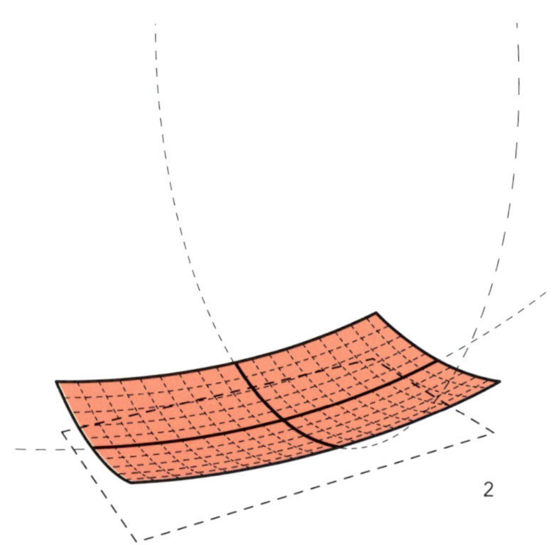

2

FORM-FINDING AND COMPUTATIONAL APPROACH

Reliance in the past on early physical hanging chain and fabric models meant that human design skills and dedication (as demonstrated by the design team who accompanied the vision of Frei Otto) were the only available tools. They made it possible to create the various measurement models that ultimately led to the unique cable-net design of the roof of the Munich Olympic Stadium in 1972 (Figure 5). Nevertheless, design and form-finding was a tedious exercise, and only a limited amount of complexity could be achieved in the shape.

Today, sophisticated digital computational tools assist the design of large tensile structures, creating unique forms that would have previously been considered impossible. New digital tools and physical simulations, such as wind tunnel tests, have allowed designers to make better use of material and produce lighter, thinner structures with accurately form-found geometries. Sometimes the lightness comes at a cost – the risk of dynamic instability or vibration – and this pushes the structural engineer to develop even more complex analysis to find the best shape to withstand it.

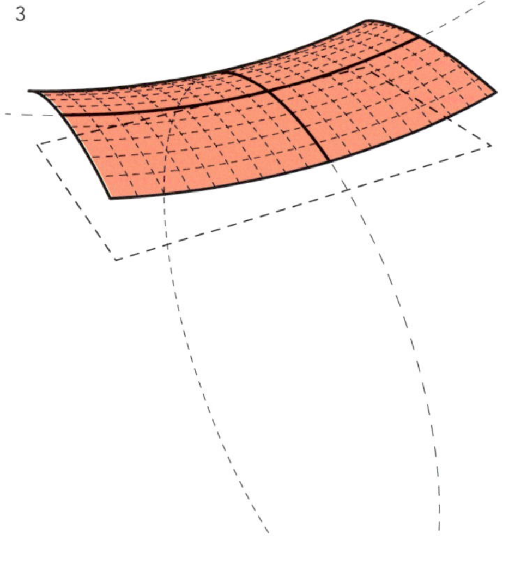

3

In digital computational form-finding, there are two main methods used to achieve the same result: dynamic relaxation and the force density method. Dynamic relaxation is a numerical method that discretises a structure as a series of simple springs, masses and forces, which are not initially in equilibrium. The counterintuitive trick to finding the equilibrium problem is to simulate the structure oscillating. If damping is then applied to this motion, the solution of the static problem becomes the limiting equilibrium position of the damped dynamic motion.

The force density method is based on a mathematical solution of the non-linear equilibrium equations for the structure, which are based on the relationship between the length of the structural elements, their internal forces and the position of the nodes. If the ratio of tension force to length in each element is assumed constant – ensuring a uniform distribution of stiffness across the structure – the problem reduces to the solution of a system of linear equations, which can easily be solved with a computer algorithm.

ANALYSIS/DESIGN

Today, a large number of powerful finite element analysis software packages perform non-linear dynamic analyses that simulate the behaviour of tensile structures with their large deflections. Generally, these non-linear analyses use an externally form-found geometry as an input, although some finite element software packages are starting to include built-in form-finding routines.

For the analysis, it is important to pay particular attention to wind action, since tensile structures are particularly sensitive to its effect: because of their lightweight nature and dynamic behaviour under turbulent effects, wind often governs the structural design. Conventional structures are generally designed for vertical loads, since these are the predominant forces affecting the structure. Tensile structures, however, are light structures where the vertical loads applied (including self-weight) can be around a tenth of the magnitude of the wind force. For this reason, wind analysis has to be thorough, considering not only the constant pressure of the wind, but the pressure variation due to turbulences. Despite being random, turbulences can potentially excite the natural frequency of the structure, leading to a resonant effect, and ultimately to the collapse of the structure. Therefore, an understanding of fluctuating wind effects on tensile structures through non-linear time-history analyses is key to avoiding these undesired resonant effects.

CASE STUDY: CATENARY ROOF

Essentially, catenary structures subject to vertical loads have the advantage of working exclusively in tension which, together with minimal shapes, ensures that each element works to its full material strength capacity, allowing the structure to be extremely light and slender. However, if any of the members in a tensile structure undergo load reversal, under particular asymmetric or upward wind or seismic loading conditions, their inability to resist compression forces renders the structure locally ineffective. This in turn modifies the load pattern in the structure and leads to a potential propagation of this effect across the structure.

A synclastic cable-net (Figure 10) can be designed as a really slender and lightweight structure, spanning long distances. The stiffness of the cable (proportional to the internal tension) will rely solely on the total weight of the roof. As it behaves as a catenary, all loads are transferred through axial tension to the supports when the structure is subject to gravity alone. However, if the wind uplift pressure becomes higher than the weight of the roof, the structure becomes unstable and will suffer large deformations before finding a new equilibrium. The same applies to non-balanced loads, or horizontal actions such as earthquakes. The nature of

2 AKT II, synclastic curvature in a suspended catenary, characterised by positive Gaussian curvature.

3 AKT II, anticlastic curvature in a pre-stressed tensile structure, characterised by negative Gaussian curvature.

4

5

4 Alvaro Siza's Portuguese National Pavilion,
Expo 1998, Lisbon, Portugal, 1998.
A synclastic suspended, reinforced-concrete
catenary structure.

5 Frei Otto and Günter Behnisch, aerial view of
Munich Stadium, Germany, 1972.
An anticlastic pre-stressed cable-net structure.

14 TENSILE STRUCTURES 198-199

6 Antoni Gaudí, Sagrada Família, Barcelona,
Spain.
Physical hanging chain model.

the structure makes it vulnerable to loads other than gravity. A resilient approach would be to design a catenary structure to resist the gravity loads, with some spare capacity to resist bending and compression for non-gravity actions.

Elegant examples of mastering materials with form were developed in the Expo '98 Portuguese National Pavilion designed by Alvaro Siza with the engineering expertise of Cecil Balmond. Here, a large plaza was shaded by a pioneer technological pre-stressed concrete synclastic roof, which uses its weight to prevent movement from uplift. A different concept is found in the anticlastic cable-net designed for the Munich Olympic Games (1972), where lightness and pre-tension work together in a delicate balance in which the curvature is the driver.

Today, through more sophisticated form-finding tools and a better understanding of materials and the dynamic behaviour of structure under non-gravity forces, work on these typologies has improved, leading to the proposal of new ones. Grid-nets (Figure 12) are tensile structures constructed from the layering of steel bars into simply bolted rigid networks, avoiding complex welded fabrications and simplifying assembly. Grid-nets are also upgradable typologies, where replacing steel bars with carbon-fibre strips should allow the creation of even lighter and more economic structures in the near future. Grid-nets may define a new typology of envelope form never explored before. While still predominantly tensile, and therefore structurally lightweight and efficient, grid-nets may not rely on an anticlastic form like a cable-net as they display a small bending capacity necessary to withstand dynamic loadings. Synclastic forms can be easily

7 AKT II, section of an anticlastic pre-stressed cable-net or membrane structure.

8 AKT II, analysis of reaction forces at the base of a cable-net structure where perimeter anchor forces at primary cable locations are larger due to global lateral stability loads.

7

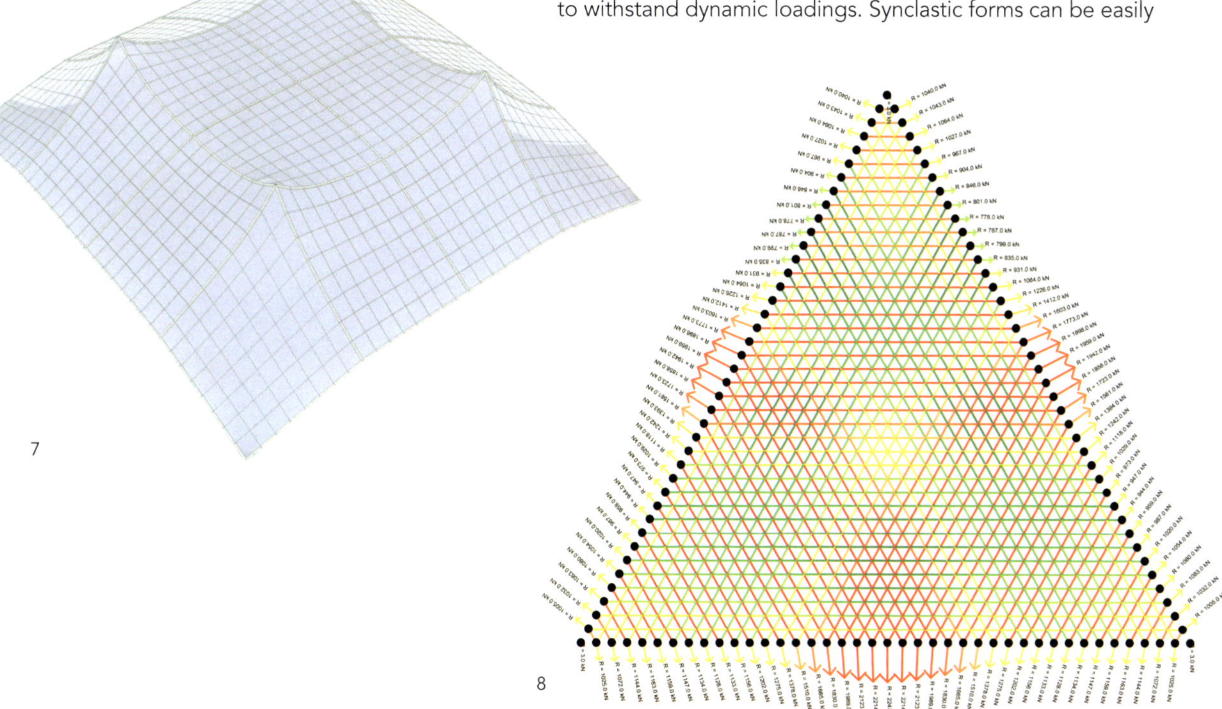

8

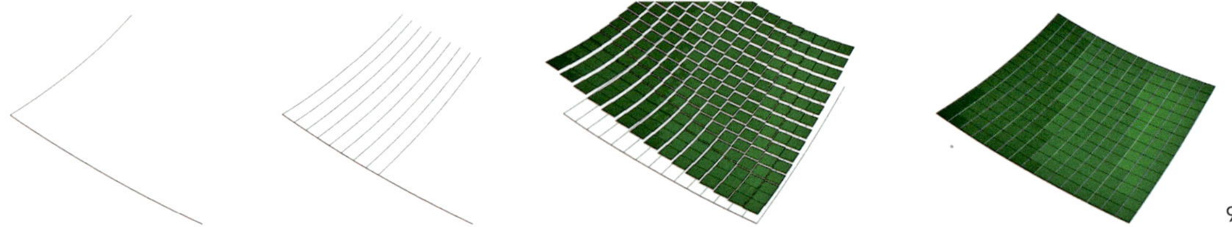

9 AKT II, flat panelisation of a synclastic translational surface using a translational surface method.

10 AKT II, geometric warp analysis in cladding panels of an anticlastic cable-net structure.

11 AKT II, typical detail of cable-net structure.

12 AKT II, typical detail of a grid-net structure.

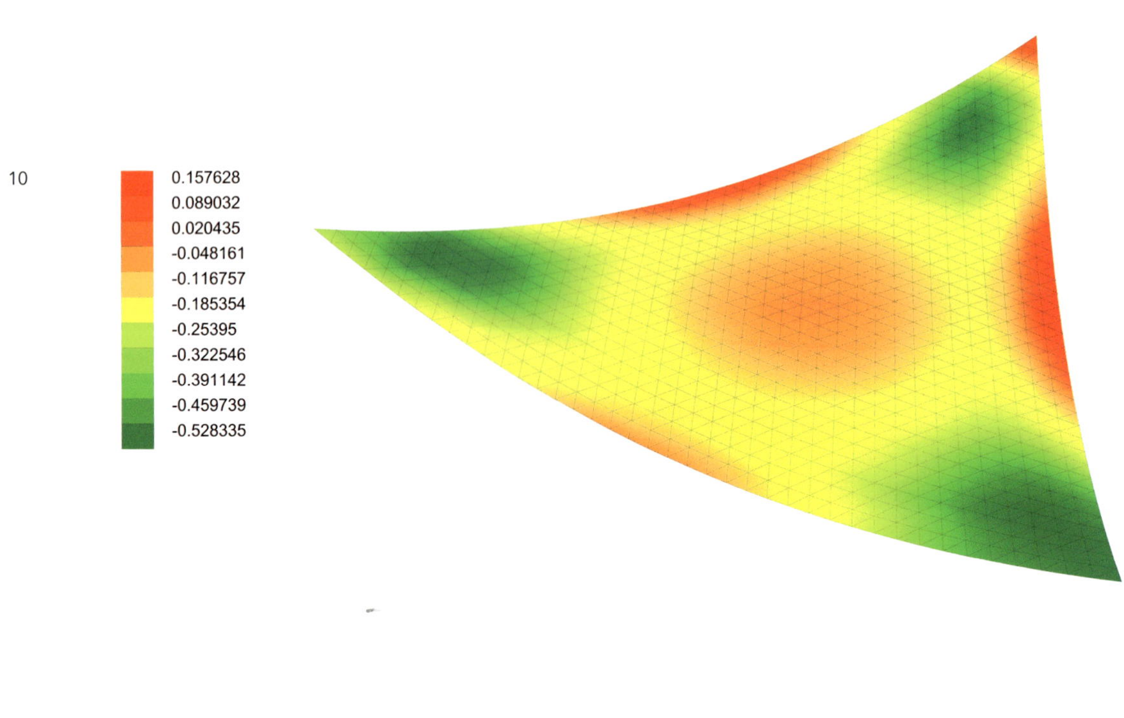

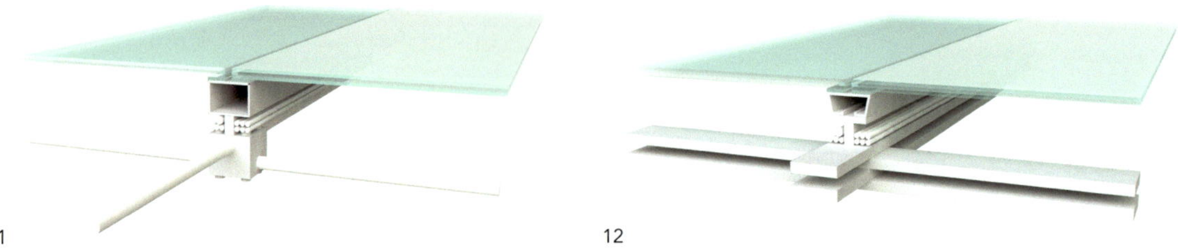

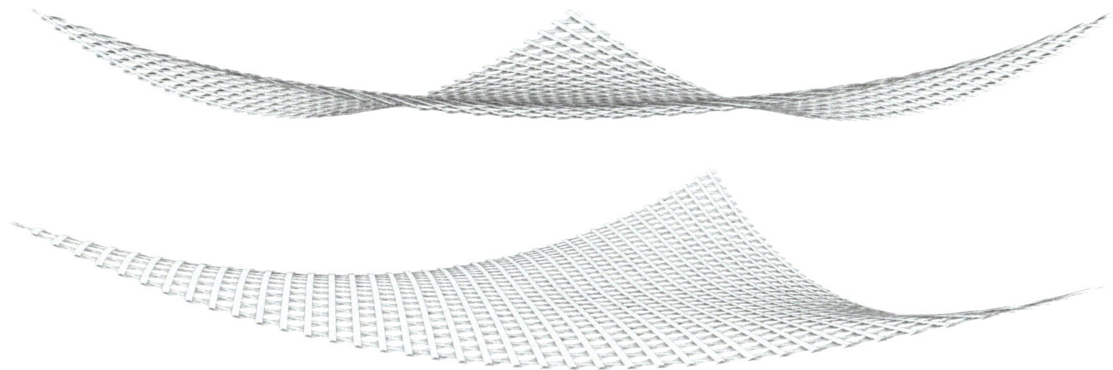

13

13 AKT II, study for a hybrid grid-net structure, composed of slender layered steel sheets or bars.

rationalised into fully planarised quadratic cladding panels, which are considerably cheaper to fabricate and install than hot or cold formed panels.

CONCLUSION
With the increase of material research, technological evolution in construction methods, and advancement of form-finding and optimisation tools, the future design engineer needs to respond with hybrid solutions which holistically combine architecture, structure and environmental fields into a precise performance, one set to higher efficient and sustainable parameters. Engineering exploration of more innovative solutions, based on the theoretical and practical knowledge of the past, will support a better understanding of the behaviour of structures which will lead to an increasing use of tensile structures. In contrast to early expectations of cable roof demand, this use is relatively modest, due to the fact that architects and engineers are generally not familiar with this sophisticated design, and are reluctant to consider such solutions appropriate for long-span roofs.

14 AKT II, CFD wind analysis on a long-span lightweight structure to assess pressures on building enclosure.

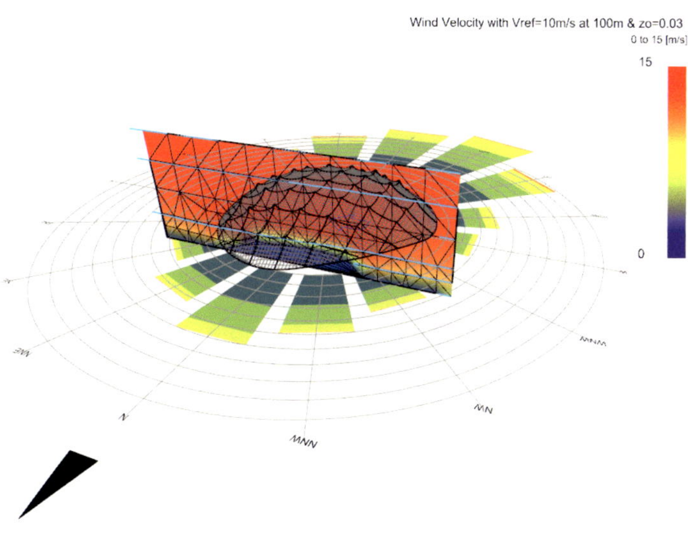

14

15

16

AIR-SUPPORTED STRUCTURES
James Kingman

HISTORY

In the phylogenetic tree of structural typologies, those described as air-supported are a recent branch of the broad and deep-rooted family of tensile, pneumatic structures. In general, pneumatic structures are characterised by their reliance on pneumatic action for pre-stress rather than the more typical mechanical means. In a structural context, the term 'inflatable' evokes images of sausage-like 'air beams' evident in many prototypical and temporary settings. Unlike these air-inflated structures, air-supported structures exhibit a unique characteristic: the pneumatic volume also serves as the occupied space. Despite the deep roots of the tensile, pneumatic typology, the history of the air-supported structure is limited to no more than several decades.

15 Davis, Brody, Chermayeff, Geismar, deHarak Associates / David Geiger–Horst Berger, US Pavilion for the 1970 Expo, Osaka, Japan, 1970.
The use of a cable-net to strengthen an air-supported membrane enabled a clear span of 460 ft to be realised in a region subject to typhoons and earthquakes.

16 O'Dell, Hewlett & Luckenbach / Geiger Berger Associates, Silverdome, Michigan, USA, 1975.
The Silverdome in use as an American football stadium. The spaces created beneath air-supported canopies were extremely flexible due to their long span, column-free nature making them ideal for hosting large-scale events in inclement weather conditions.

Requiring a lightweight, deployable enclosure, opaque to the elements, engineers crossbred the millennia-old concept of the frame-supported tent with the Victorian curiosity for lighter-than-air balloons. The resulting spheres, or 'radomes', each formed of a single layer structural membrane, are the genesis of the air-supported typology. Requiring a 100,000 ft^2 (9.3 m^2) unobstructed enclosure for the US Pavilion at the 1970 World Expo, in the seismically and meteorologically hazardous city of Osaka, David Geiger took the concept of the radome further by merging it with contemporary advances in cable-net design to create what remains, even today, the world's lightest long-span enclosure.

SUSTAINABILITY

Early explorations in permanency and scale seemingly encountered the paradox that as the self-weight of a structure tends toward ever lower values of 'efficiency', the euphemistic nom de guerre of cost fails to scale accordingly. This phenomenon is thrown into sharp focus in the case of air-supported structures, where the extremely light physical and capital cost is overshadowed by a hidden cost: energy.

The changing sustainability zeitgeist and a series of spectacular collapses (Figures 17 and 18) caused even the pioneers of air-supported structures to reassess their fitness in permanent and large-scale applications. These factors, and a tendency for some engineers and architects to take the theoretically limitless spanning capability of air-supported envelopes and stray into the absurd, have created an image of the typology as a retro-futuristic folly, derided as wholly inappropriate for serious consideration in a 21st-century context.

17 Skidmore, Owings & Merrill / Geiger Berger
Associates, Herbert H Humphrey Metrodome,
Minneapolis, USA, 2013.
Deflation of the air-supported canopy.
Although highly stable under strong winds, the
imperceptible pressure differential stabilising an
air-supported canopy can be overcome by even
small build-ups of snow.

18 O'Dell, Hewlett & Luckenbach / Geiger
Berger Associates, Silverdome, Michigan, USA,
2013.
Deflation and damage to the Silverdome
followed a period of neglect. Without a
constant supply of energy in the form of
pressurised air, an air-supported structure is
susceptible to irreparable damage.

17

18

Adaptability, however, is an increasingly desirable trait in architecture. Whether considering technology-based organisations with future uncertainty regarding office space requirements, or an immediate response to a humanitarian crisis in an increasingly unpredictable climate, a degree of flexibility is required that is simply not possible with a traditional building envelope. The extremely low capital cost of air-supported envelopes, compared with rigid alternatives, offers the unique possibility of simply removing and replacing the canopy, economically, in a relatively short time (Figure 20).

There is a constant demand for energy simply to maintain the form of the structure. However, in most practical cases a demand also exists to maintain the required number of air changes for occupancy comfort, a number which comfortably covers the requirements for pressure loss or leakage through the air-supported enclosure. Rapidly evolving technologies for creating renewable energy, such as flexible solar panels, can be integrated within the inflatable envelope to further mitigate this energy usage. It is entirely possible that, even looking at a relatively short time horizon, such technology may be sufficiently developed as to make such an air-supported structure self-sustaining.

19 AKT II, scaling of air-supported structures.
Typically the weight per unit area of rigid
spanning enclosures will square with their span.
Air-supported structures are unique due to their
ability to scale without any significant increase
in self-weight per unit area.

20 AKT II, cost versus time of an air-supported
structure compared with a typical spanning
enclosure.
Air-supported structures are unique in that they
require a constant supply of energy in the form
of electricity to power fans simply to maintain
their form.

19

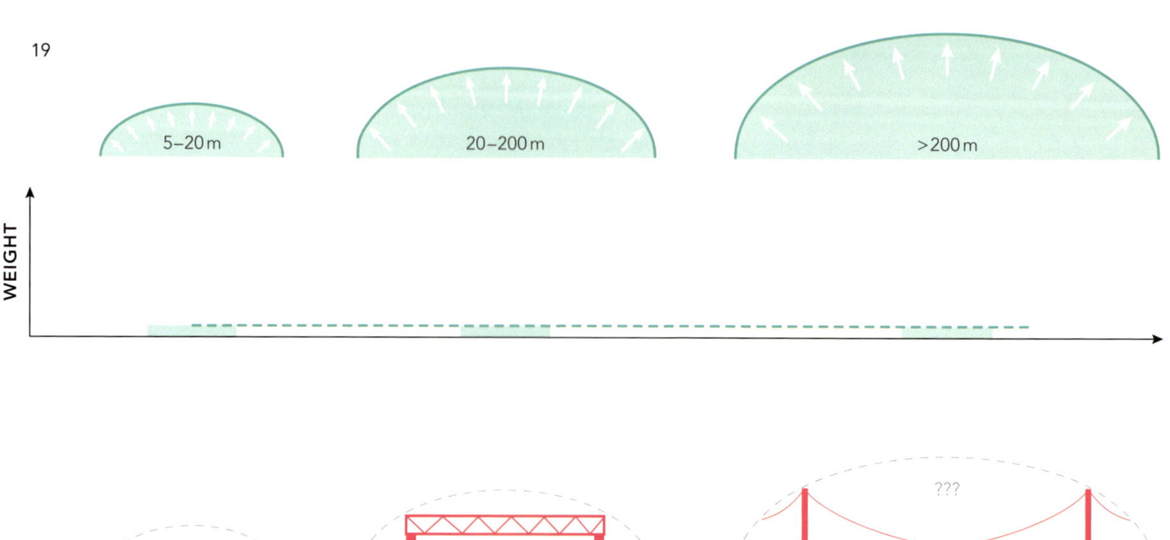

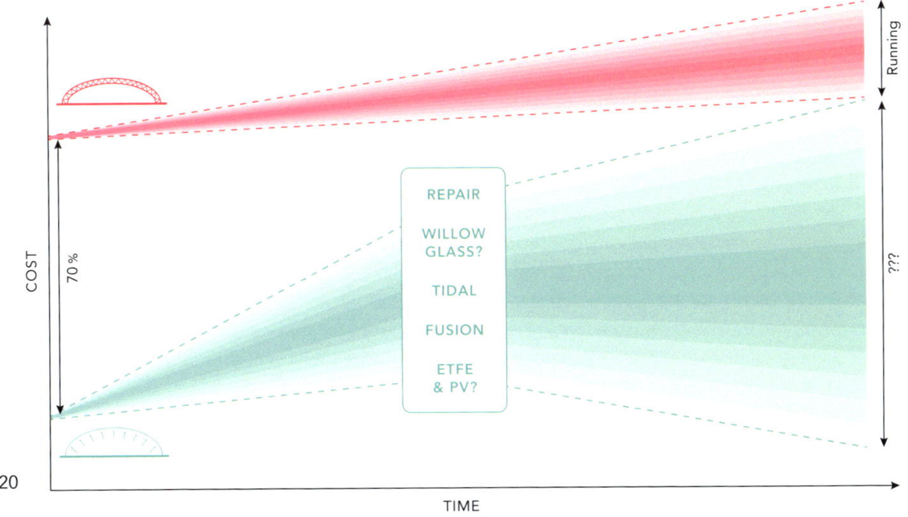

REDISCOVERY

The dominance of gravitational forces acting upon prosaic structural typologies for spanning enclosures, such as beams or trusses, means that their mass will typically scale according to the square of their span. Air-supported structures do not scale according to such fundamental rules; practical difficulties associated with creating seals and ensuring occupancy comfort place an upper limit on the self-weight of air-supported structures in the order of 10–20 kg/m^2 (Figure 19). By using air pressure to step outside the constraint of gravity, the scope of form that can be efficiently realised with an air-supported structure is that of synclastically curved, circular geometries rather than the typical anticlastically curved catenary geometries of mechanically pre-stressed tensile structures.

The profile of the US Pavilion, and all large-span air-supported structures, is sculpted such that the air flowing over the structure is deflected by a perimeter barrier and, just as the airflow over the wing of an aeroplane pulls it upwards, the air flowing over the canopy pulls it outwards, stiffening it in the process. The faster the wind blows over the structure, the stiffer and less prone to deformation the canopy becomes, thus making air-supported structures almost unique in their ability to passively adjust their stiffness in response to environmental loading.

At the nanoscopic scale, the problem of determining the response of an air-supported structure to wind loading can be conceptualised as determining the response of a thin membrane to bombardment by a series of air molecules of different energies from either side: the membrane will tend to move away from the higher energies towards the lower energy region. Also, the action of the membrane itself moving will either add or remove energy from this system, leading to an ephemeral balance of forces. Today, parametric design codes and relatively rapid computational fluid dynamic (CFD) analysis have increased the level of certainty surrounding the environmental loading acting upon an air-supported membrane, enabling the design of increasingly ambitious forms. Calculating the response of the structure to such action, however, remains challenging.

Historically, a pragmatic engineering approach, using the governing equilibrium equation of a synclastic form (pressure x radius = force) has been the limit of analytical capability in relation to tensile structures for all but a specialist cabal of nerds. As is typical for structures with a self-weight significantly less than imposed loading, deformations are generally the governing design criteria; as they cannot readily be assessed by analytical means, numerical techniques are necessary. Numerical analysis for the modern structural engineer is the creation and solution of finite element models which are almost exclusively formulated using the global stiffness matrix approach. Embedded within this is a fundamental assumption of small displacements, making its application to the response of air-supported structure problematic.

21 AKT II, designing without the constraint of gravity.
Parametric definitions of codified wind loading combined with dynamic relaxation engine Kangaroo Physics enable real-time simulation of wind loading.

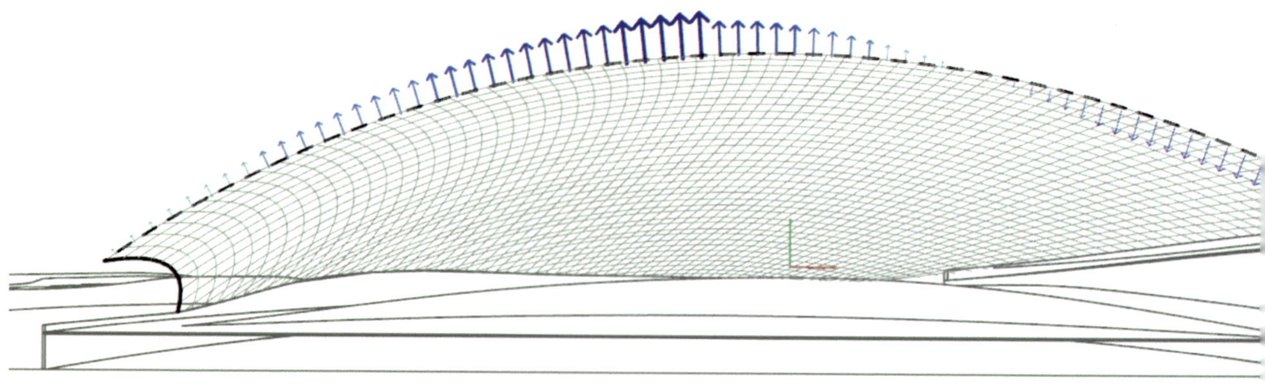

Wind Pressure Vectors

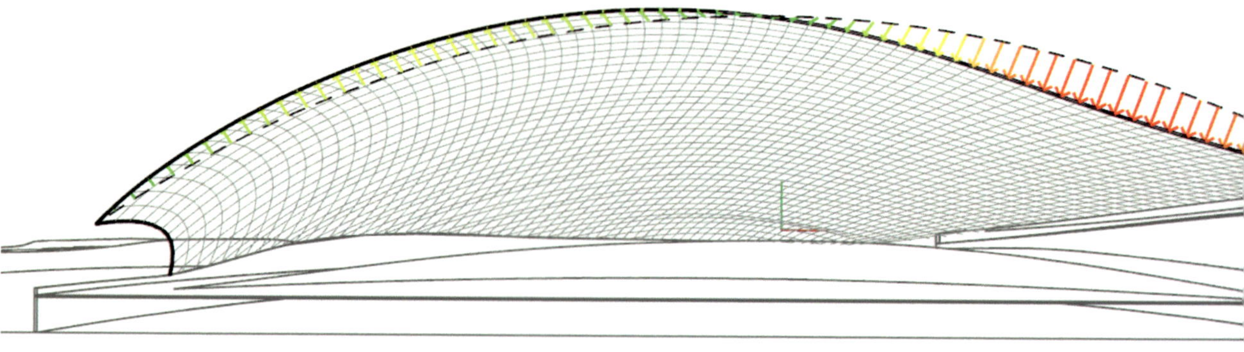

Deformed Dome

Higher order finite elements and iterative solution procedures must be employed, leading to a degree of conceptual and mathematical complexity unsuited to early stage design. SAP2000 and SOFiSTiK, the de facto analysis packages in modern practice, have proved extremely ill-suited to the purpose. A typology as youthful as the air-supported structure requires a new approach.

Unlike the structural engineer, obsessing over the last decimal point of accuracy in the simulation of structural behaviour, the computer graphics (and particularly the game and app) design community has rapidly developed the capability of creating accurate, real-time simulations of complex physical phenomena. In our arrogance, all but a few engineers have failed to recognise that at the heart of everything, from *Angry Birds* to *Call of Duty*, simulating a complex environment in real-time is an algorithm originally developed for structural analysis: dynamic relaxation.

By combining a custom-scripted parametric description of the wind loading acting upon the structure with a dynamic relaxation engine in the form of Kangaroo Physics, it is possible to create an almost real-time simulation of the complex behaviour of even large air-supported structures.

MATERIALITY
Over the past few decades ETFE, in the form of a stressed membrane or air-inflated cushions, has been used extensively in cladding applications as an alternative to traditional glass panels. An air-supported envelope formed of ETFE is an obvious avenue for innovation. The relatively low strength and fragility of commercially available ETFE makes any large-scale air-supported envelope formed solely of a single sheet of the material practically impossible. ETFE cushions, used as infill panels for a steel cable-net, solve this problem and offer a degree of resilience if a single layer of ETFE fails. Multi-chamber ETFE cushions also enable control over the pressure in each of the chambers independently of the pressure within the occupied space (Figure 22).

The strength and robustness limitations of ETFE today remain a significant barrier to the implementation of the material in a large-span envelope. ETFE manufacturers are currently investigating the possibility of laminating the material with microscopic reinforcing strands of tensile fibres to create a material which is both strong and stiff. The development of these composite materials is not yet sufficient to even begin testing at a prototypical scale, but the future possibilities of forming air-supported structures from such a material are promising.

THE FUTURE
Despite significant promise and moderate popularity during the middle of the 20th century, the air-supported typology faded from the conscience of the engineering and design communities over the following decades. But now, an increasing demand for adaptable and flexible architecture necessitated by, for example, changing demographics, workforce uncertainty and climate change, makes

22 AKT II, structural concepts for air-supported
structures utilising an ETFE membrane.
A significant barrier to the implementation of
ETFE in air-supported structures is the fragility
of the material.

23 AKT II, critical issues to the design of
air-supported structures are air loss and
redundancy.
Strategies for providing inherent redundancy
to air-supported structures rely on the catenary
action of the membrane in the deflated
condition.

24 AKT II, applications for air-supported
structures in an increasingly uncertain world.
The extremely low capital cost of air-supported
structures is increasingly relevant where
the certainty regarding the future use of a
space may only extend to five years or where
an immediate response is required to a
humanitarian crisis.

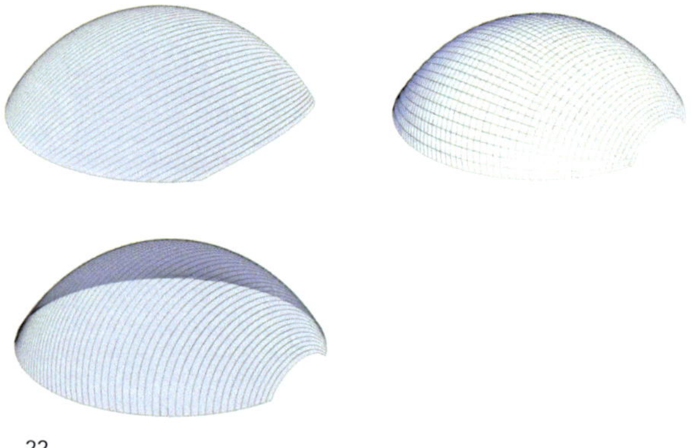

22

23

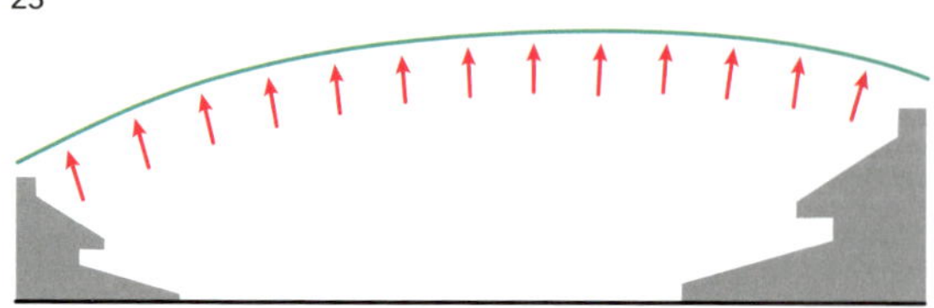

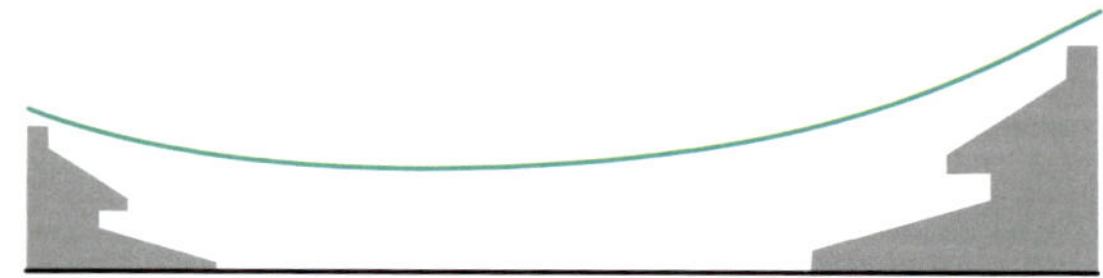

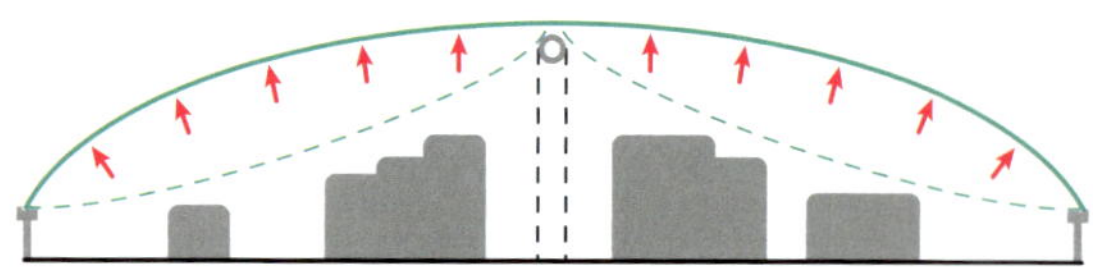

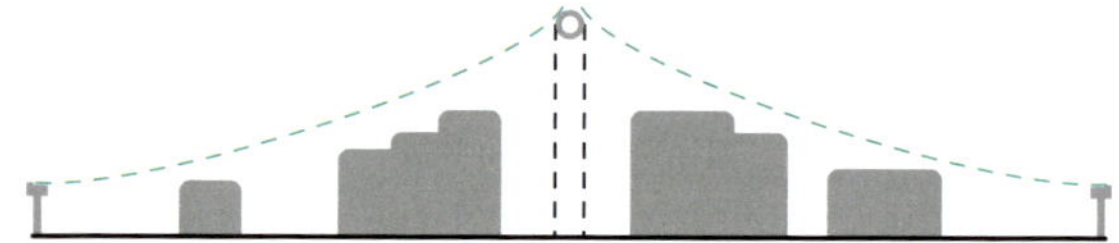

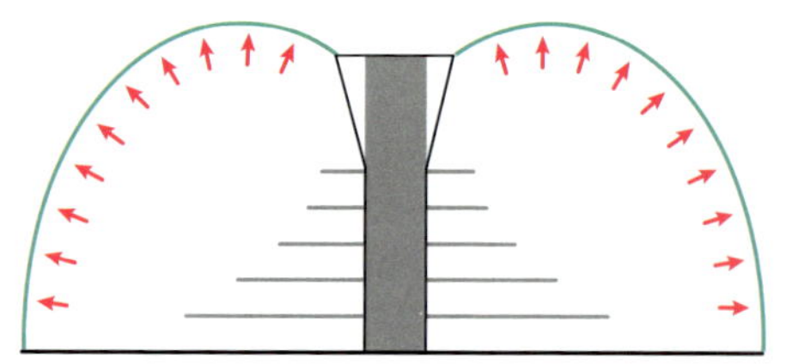

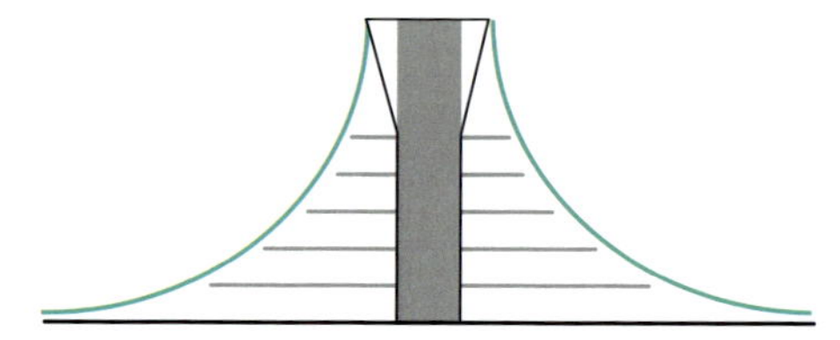

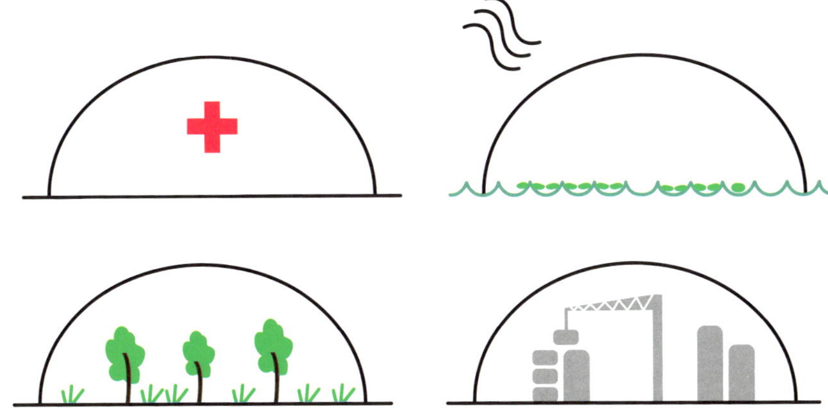

a contemporary review of air-supported structures fitting. A rediscovery of the structural principles of the typology is needed; unparalleled economy of material and massive scalability make it desirable to understand how far the typology can be developed within today's landscape of materiality and analytical tools.

A parametric description of wind loading acting upon an air-supported membrane and a dynamic relaxation-based analysis engine, all developed using free plug-ins and custom scripts within the Rhino-Grasshopper environment, has enabled a complete workflow of form-finding, analysis and design to be performed in almost real-time. Such a freely available and powerful design tool makes it possible for a wider spectrum of engineers and designers to develop the air-supported typology.

REFERENCES

1 Robert Hooke, *A Description of Helioscopes and Some Other Instruments* (London), 1675.

TEXT

IMAGES

1 AKT II, 2015; p.art atmosphere.
The environments in which AKT II's applied
research team navigates, builds connections
and plays an active role whether the goal is the
delivery of a design or speculative research.

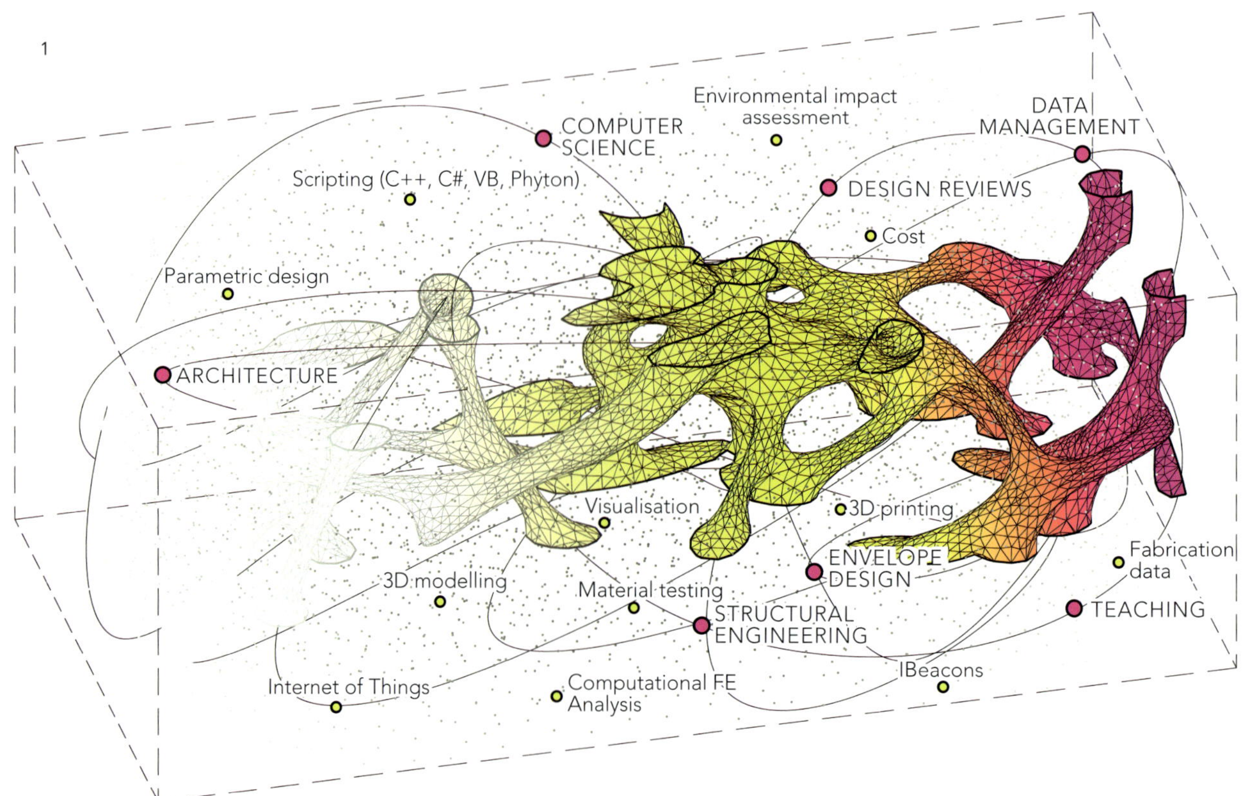

1

15 INTERWEAVING PRACTICE

EDOARDO TIBUZZI

Tools are extensions of the human mind, developed to perform a task and achieve goals. They are physical or digital 'artefacts' which bridge intentionality and execution. Crafted by human ingenuity and skills, they are designed to satisfy the basic human needs for survival, well-being and aspiration. From the beginning of humanity, when sharp rocks were used to carve and hunt, to the present day, when smart phones have drastically changed the way we communicate and interact, we have been constantly designing and refining tools to perform tasks better and faster. Today, more than ever, acceleration in technological advancements and an increase in the need for more efficient use of resources have made us question what drives the effective development of tools. Ideas, resources and time play a fundamental role in their development. The more ambitious a vision, the more resources or time are required to achieve it. Resources include capital, material and skilled people; the right balance of these resources and time ensures that a good concept is implemented on time and on budget, while achieving its goal.

Today, in the world of design, the development of digital tools has changed considerably, rapidly evolving over the past couple of decades. The notion of computer-aided design (CAD), for example, has transformed from its original 1960s idea, from 'design tool' to 'drafting system'. Whether this has been the result of economic market pressures, or because the time wasn't quite right, the idea of creating a means to enhance the creativity of the designer was somehow replaced by the idea of providing an efficient method of drawing production. Before that original CAD acronym was coined, interactive computer graphics were exclusively the subject of science fiction. It was a time when computers were operated with pre-punched cards, such as the Whirlwind at MIT, so the technological gap between the idea and realisation of computer-aided design was huge. Investment in the development of computer hardware and software has filled that gap in a relatively short period, although new concepts – such as the emerging community of 'computational designers' – have perhaps come closer than producers of commercial CAD software to fulfilling the original promise of creating a 'tool for design'.

The evolution of hardware saw the 250-ton Whirlwind, powered by 12,500 vacuum tubes, filling the equivalent of a two-storey

2

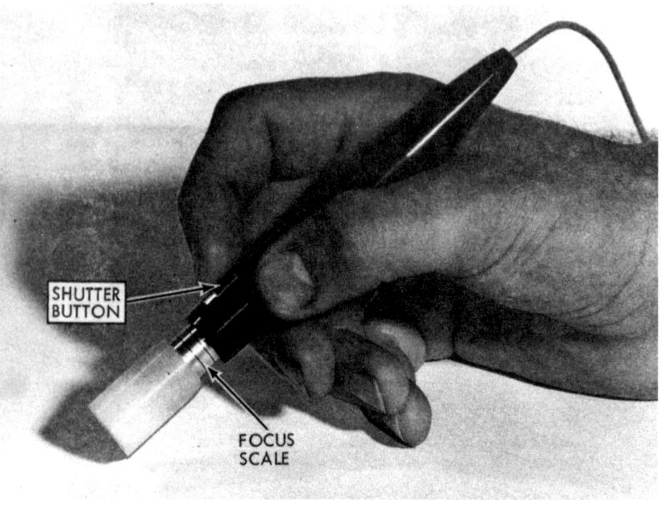

3

2 Ivan Sutherland, 1963; Sketchpad: A Man-Machine Graphical Communication System.
The pioneering work of Sutherland, on enabling computer and computation programs to interpret information drawn directly on a computer display, paved the way for the CAD/BIM revolution.

3 Ivan Sutherland, 1963; Sketchpad pointing device.
The revolutionary device introduced by Sutherland allowed the computer to interpret the geometries drawn directly on a display.

house, and the IBM 5 MB hard drive, needing six men to transport it (Figure 4). It was in those years that Ivan Sutherland, a student at MIT, had a vision of the future: he imagined people using computers to design. In his doctoral thesis, *Sketchpad: A Man-Machine Graphical Communication System*,[1] he described his remarkable idea of a computer-aided design tool (Figures 2 and 3). By the 1970s many CAD software companies were founded to commercialise CAD programs and hardware. By the end of the decade, major researchers in the field started to extend CAD from 2D space to 3D environments. More sophisticated models of 3D objects were introduced by the work of Ken Versprille who, in his PhD thesis, proposed NURBS as a simple way of creating complex surfaces. The 1990s saw the release of several 3D modelling kernels, most notably ACIS[2] and 3D CAD Parasolids. By the end of the decade, the power of PC hardware caught up with the computational requirements of 3D CAD.

Today, Autodesk has managed to establish itself as a leader in the CAD market, first with the development of AutoCAD[3] and then with its BIM[4] counterpart, Revit.[5] These programs predominantly specialise in the production of 2D drawings and 3D models during the detailed documentation phase of a building project. Most CAD packages still lack effective ways of interfacing with analysis packages. Their use is therefore mostly limited to visualisation and geometric coordination. Some programs, such as Rhino, have developed programmable parametric platforms such as Grasshopper, which allows the interface with external analysis software, including FEA and CFD[6] or other bespoke software packages developed by computational designers, to execute specific functions.

In recent years, the more advanced development of design software has taken place on open-source platforms, which have allowed analytical solvers and bespoke custom applications – developed by designers themselves – to interface with project-specific parametric definitions. This open and flexible approach to computational design has allowed the sharing of specialised software, geometric definitions, structural and environmental analysis tools, and data between designers, consultants and contractors across the industry. Access to shared online databases of information, such as custom-building components (eg, standard steel profiles for structural analysis) and weather data for environmental analysis, has allowed the sharing and improvement of this data.

PACKAGE VS INTERFACE

In simplistic terms, the process that leads to the realisation of a building starts from an idea, develops through its design, and is implemented through its construction; finally, the building is managed and maintained throughout its life. With a traditional process, various parties operate in partial or total isolation from each other; architects, engineers and specialists, project managers, quantity surveyors and contractors transfer information from one to the other in a linear sequence (although inevitable, overlaps and design iterations are infrequent and considered

4

4 IBM, 1956; 305 RAMAC 5 MB hard drive. This 1956 HDD was composed of 50 24-inch disks, weighed over a ton and it was leased for about $3,200 per month. Not quite ready for today's laptops.

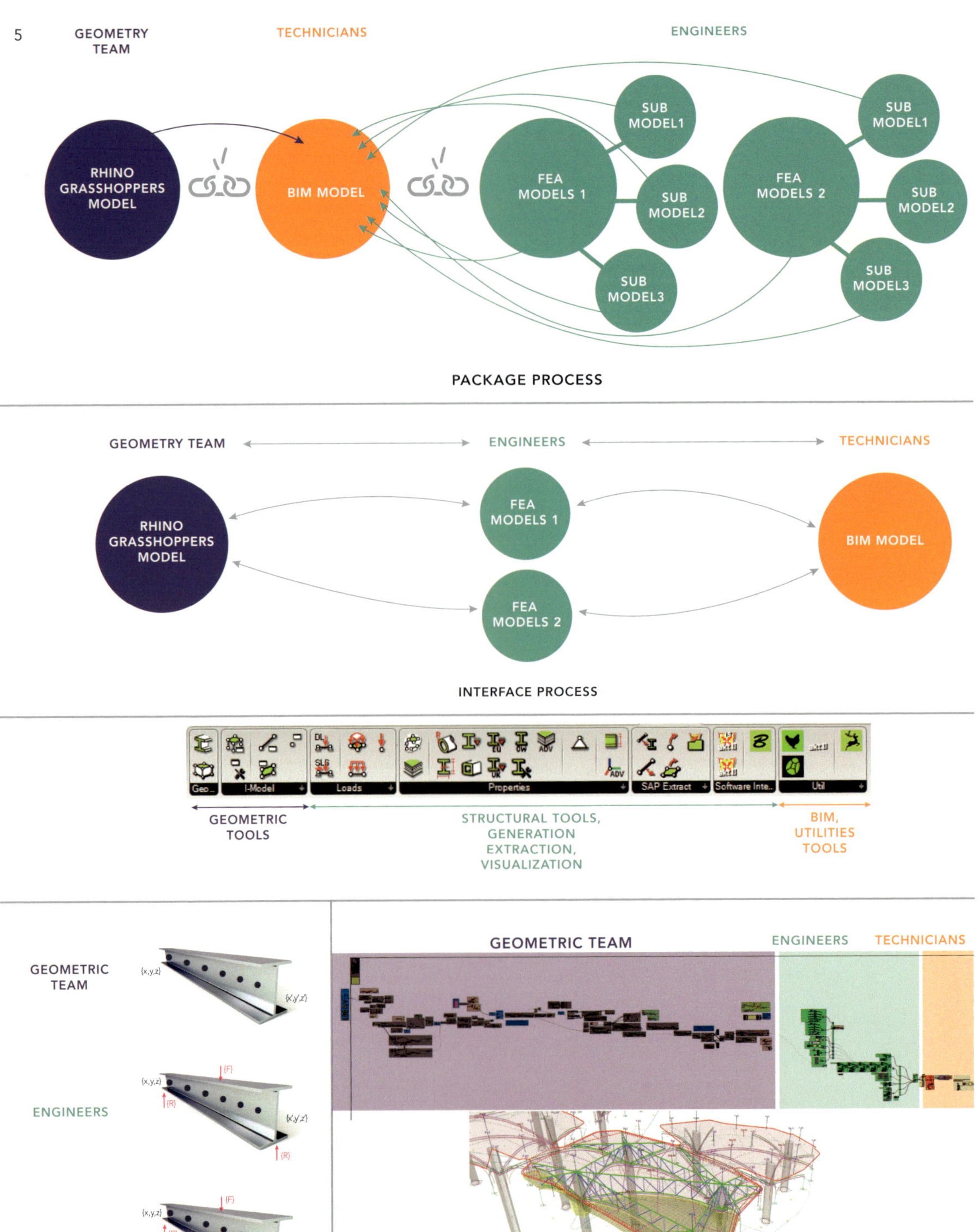
5
GEOMETRY TEAM
TECHNICIANS
ENGINEERS
RHINO GRASSHOPPERS MODEL
BIM MODEL
FEA MODELS 1
FEA MODELS 2
SUB MODEL1
SUB MODEL2
SUB MODEL3
SUB MODEL1
SUB MODEL2
SUB MODEL3
PACKAGE PROCESS
GEOMETRY TEAM
ENGINEERS
TECHNICIANS
RHINO GRASSHOPPERS MODEL
FEA MODELS 1
FEA MODELS 2
BIM MODEL
INTERFACE PROCESS
Geo...
I-Model
Loads
Properties
SAP Extract
Software Inte...
Util
GEOMETRIC TOOLS
STRUCTURAL TOOLS, GENERATION EXTRACTION, VISUALIZATION
BIM, UTILITIES TOOLS
GEOMETRIC TEAM
ENGINEERS
TECHNICIANS
GEOMETRIC TEAM
ENGINEERS
TECHNICIANS
{x,y,z}
{x,y,z}
{F}
{R}
{R}
{x,y,z}
{x,y,z}
{F}
{R}
{R}
{UB 254 x 146 x 31}

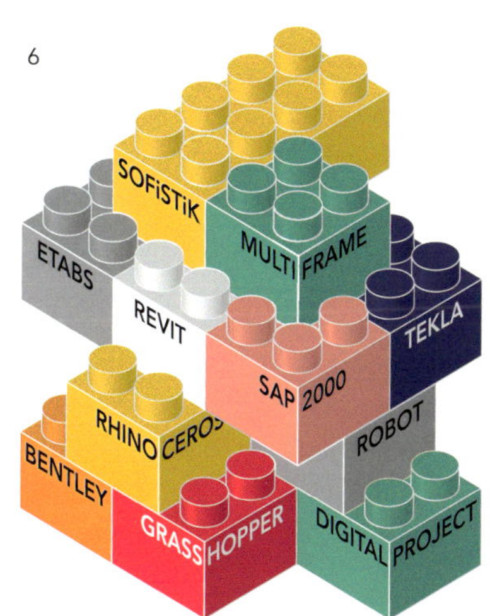

inefficient). This linear process is tried and tested and low-risk, so its outcome is more predictable and conventional.

In the traditional process, the tools used are specialised and confined to single areas of competence, developed by specialised software companies to address specific tasks; these are fundamentally 'closed boxes' producing typical pre-engineered solutions. Most of the CAD and finite element packages available today, used in the conventional way, fall under this category. Their functionality tends to be as generic as possible in order to address the widest possible range of options. They are simple and inflexible to reduce errors.

This linear, serial and constrained approach to design becomes inappropriate and inefficient when applied to a building where either the form or the integration between parts is more complex and unconventional than usual, or where a higher degree of coordination between parts is required than in a mere 2D layering of floor plans. In this instance, a more integrated iterative approach is required, one which allows for the testing of new ideas by different disciplines. At this point, an efficient, real-time exchange of design information becomes critical to the efficiency of the design process.

The past decade has witnessed the delineation of two approaches to the development and use of CAD software. The first is the integrated package, and the other is the 'interoperable' approach. The integrated package is software that comes with its own proprietary modelling interface (sometimes a parametric interface), a predefined data-structure and specific output formats (eg, Revit). Although the software has ways of importing or exporting 'dead' geometry from or to other packages, this is a cumbersome, indirect procedure that cannot be fully automated.

The interoperable model, on the other hand, is an object-oriented approach to CAD software, based on the ability to interface different software packages in a flexible, interchangeable way by accessing the functionality through their APIs.[7] The interoperable model also allows the development of custom interfaces and components that carry out specific tasks in a flexible, reconfigurable and logical framework, and which are utilised in form-finding and optimisation processes. The interoperable model uses intelligent models, 'i-models',[8] which transfer live geometry and properties, material properties, loads, boundary conditions and other data that characterise the structure. The i-model is a 'tagged model' which allows dynamic data structuring. In other words, its data can be organised and reconfigured through the development of the model without it being dictated by the software package that it inhabits; it is reconfigurable (re-groupable, re-layer-able) in different ways depending on the needs of the project.

Most BIM packages, and parametric programs like Revit, CATIA[9] and Digital Project,[10] are project execution packages, capable of carrying large amounts of heavily structured data in a very

5 AKT II, 2015; package or interface, that is the question.
The inefficient process of designing using multiple disconnected platforms was the initiator of the development of AKT II's internal Interface (Re.AKT).

6 AKT II, 2015; reunite the intelligence.
AKT II's interface allows multiple software to have a dialogue and share relevant information on the project during the design, allowing the engineers to forensically access all the proprieties available at any point in time.

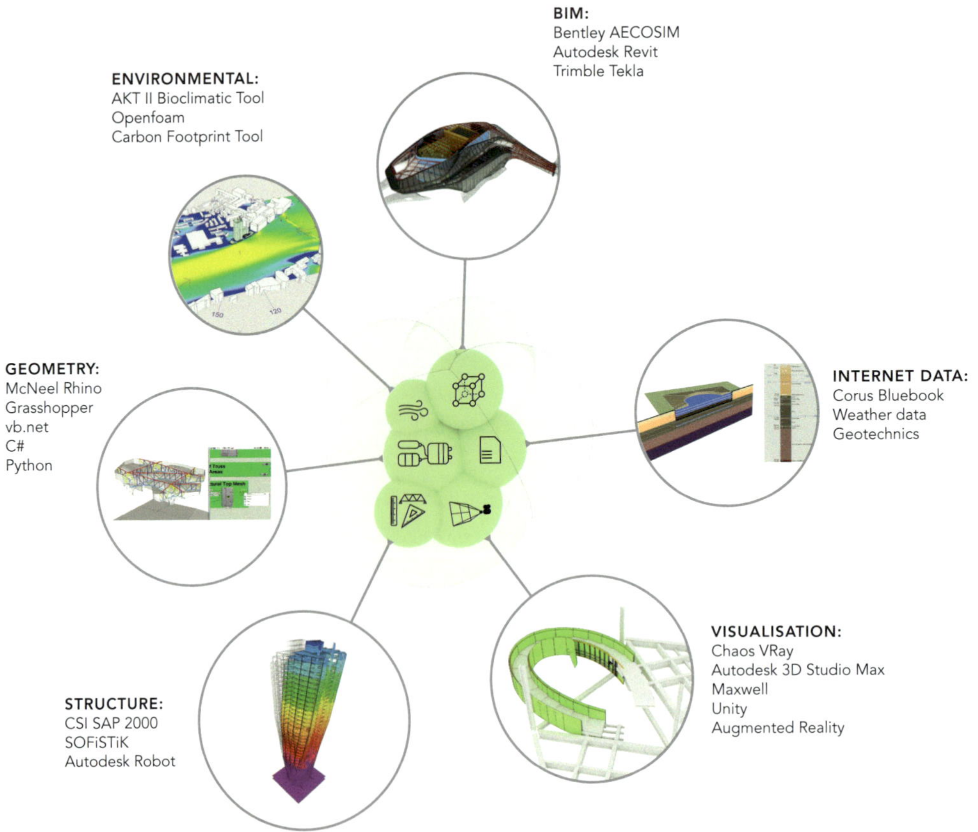

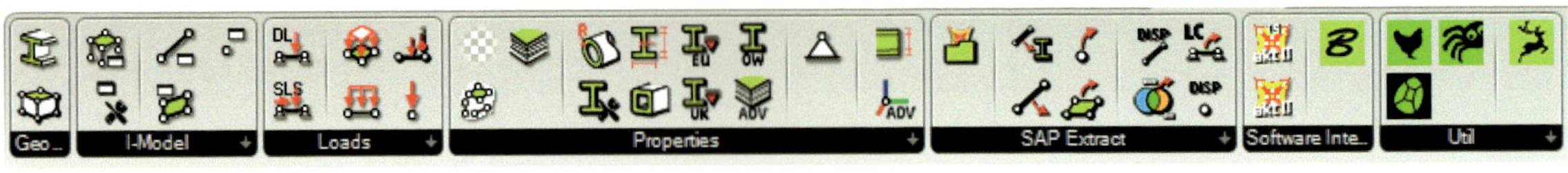

7 AKT II, 2015; Re.AKT.
AKT II interface combines geometry, structure,
environmental, BIM and Internet data in one
accessible database.

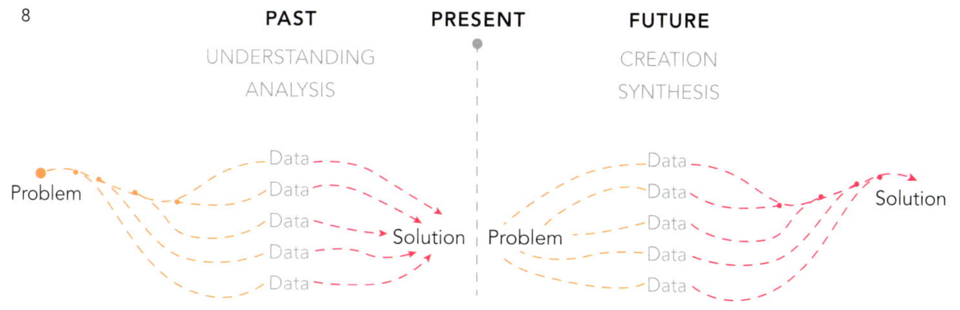

8 AKT II, 2015; blending analysis and synthesis.
Working with an interface allows us to resolve
a problem using both analysis and synthesis in
a continuous loop until the desirable solution
is found.

rigid and inflexible framework. On the other hand, the i-model
is a flexible, expandable, system that can be employed from
concept through to execution. The i-model provides a logical
framework and an interface to other customised tools designed
to solve specific aspects of a project. These include modelling
complex geometry, form-finding, dynamic structural analysis
and optimisation, extraction of results, data visualisation and
production of 3D representations and 2D construction drawings.
Driven by a parametric engine, the i-model provides a single
dynamic relational system which can be 'crafted' by hand, with
real-time visual analytical information mapped onto it, or optimised
through automatic or semi-automatic routines.

ANALYSIS VS SYNTHESIS
While the process of analysis consists of the detailed decomposition
and examination of the elements of a problem, synthesis is the
re-composition of these elements within a coordinated whole.
In other words, analysis looks into the past, trying to understand
if something that is already created works, while synthesis looks
toward the future as it combines different elements into a new
creation. On the one hand, a rigorous and forensic 'engineering'
approach is used, while on the other, the parts are assembled by the
'designer'. The non-linear iteration of the two processes allows an
informed, but holistic, approach to a design.

Re.AKT is an interface developed to integrate specialised
analytical processes within the synthetic, holistic and interoperable
model of a project. This allows the multi-parametric development
of a design with the embedded, real-time feedback of the
analysis of different aspects of a building, such as its structural or
environmental performance. A system like Re.AKT streamlines
the information from multiple disciplines into a single solution by
sharing data across several different software packages. The core
of the model is based on the geometric definition of the structure
with its internal hierarchies of primary and secondary elements,
groups and layers. Subsequently, sets of properties, including
physical characteristics (such as material properties, boundary
conditions or constraints) and external actions (such as loads), are
allied to the geometric elements of the model. These attributes
can be easily visualised in a 3D environment, as can results, which
are fed back from the analysis packages and mapped onto the
base geometry.

9 AKT II, 2015; interface loop.
Re.AKT is at the centre of the process storing,
connecting and re-distributing data from all the
sources engaged in the design.

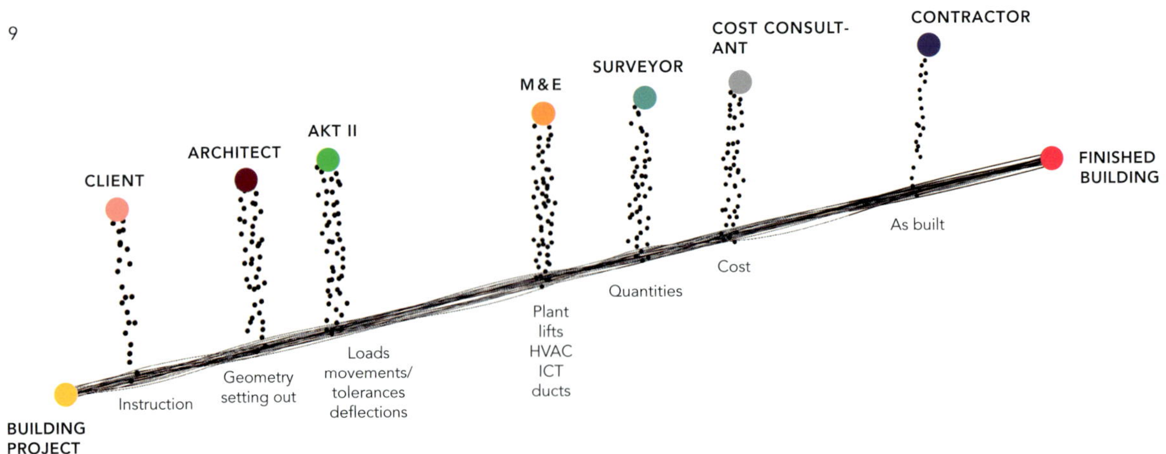

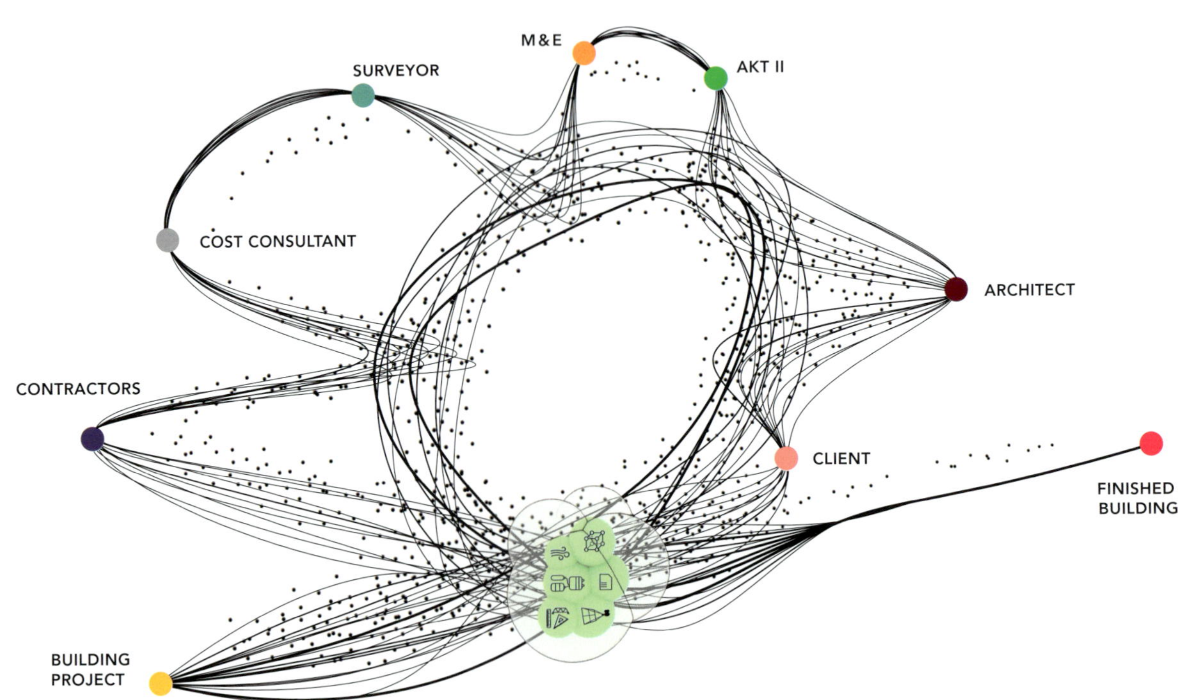

Complex structured models are constructed through the use of object-oriented interoperable design; built up gradually through the development of the design from concept through to completion, they eventually evolve into full coordination BIM models. These models can be used to automate the production of detailed fabrication information and setting out, reducing the risk of erroneous translation or over-simplification from concept-modelling software to a production package. These tools are empowering the design engineer with more control over the impact that specific changes have on the structural performance. For example, if the architectural design is modified, the seamless link to the analysis will show the areas of improvements and the areas that need redesigning. Equally, if the cost targets are decreased, the designer can address those new requirements by forensically identifying the areas to be changed without having to remodel or re-analyse the whole model.

AL FAYAH PARK

During the early stages of the Al Fayah Park project, it was quickly realised that the supporting structural skeleton was going to be directly dependent on the soffit setting out, so the route followed was a collaborative process where a geometric parametric model was developed with Heatherwick Studio. As the geometry of the soffit and its relative structural truss was generated by common parameters, so the surface evolved, and so the structural truss geometry was automatically updated. Different organisational logics were tested, as well as different vault generation methods

10 Heatherwick Studio with AKT II, Al Fayah Park, Abu Dhabi, 2015.
Al Fayah Park rendered image. Overview of the park.

10

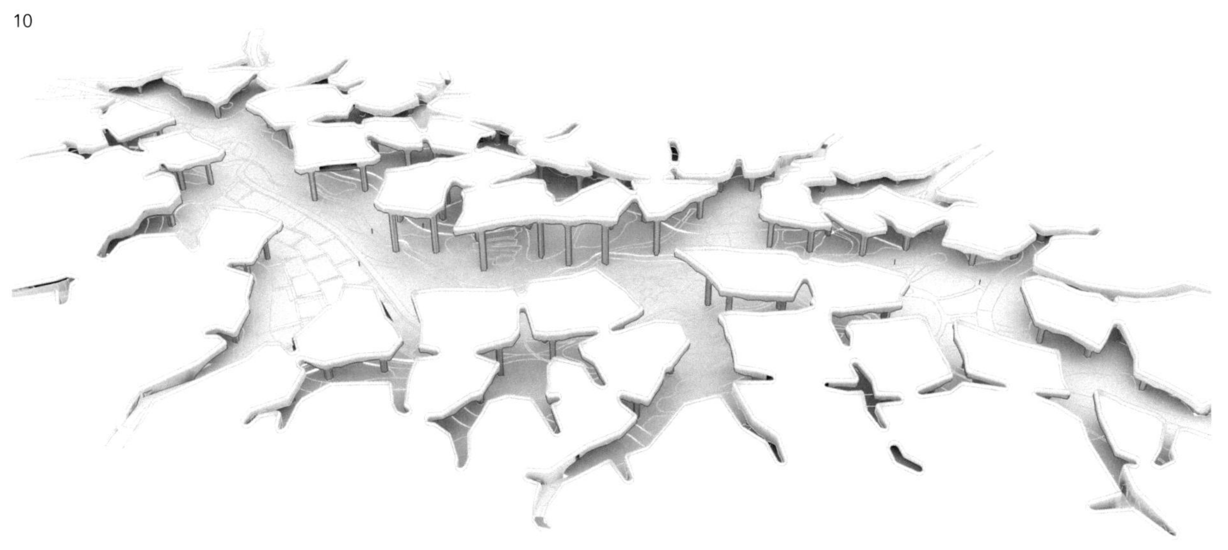

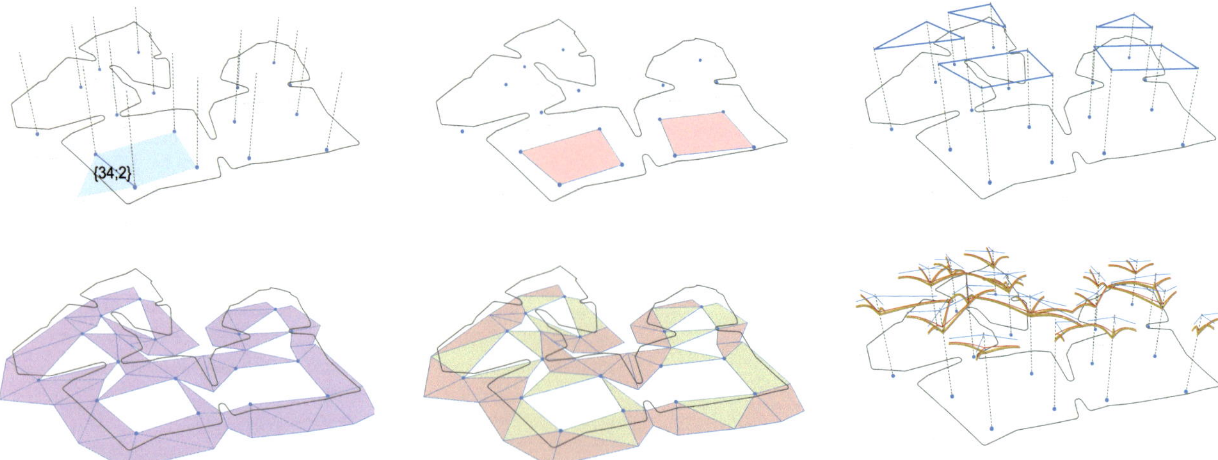

11

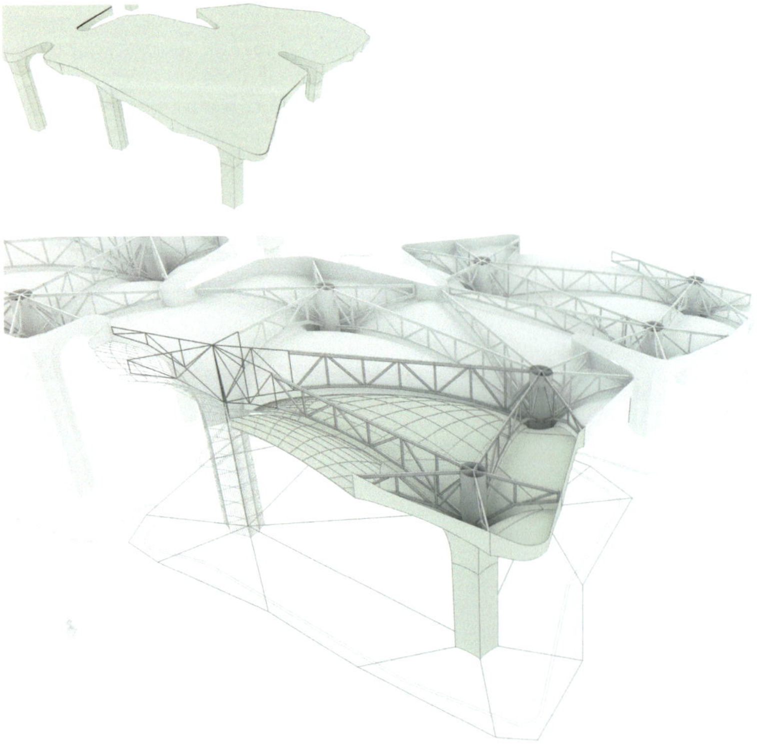

12

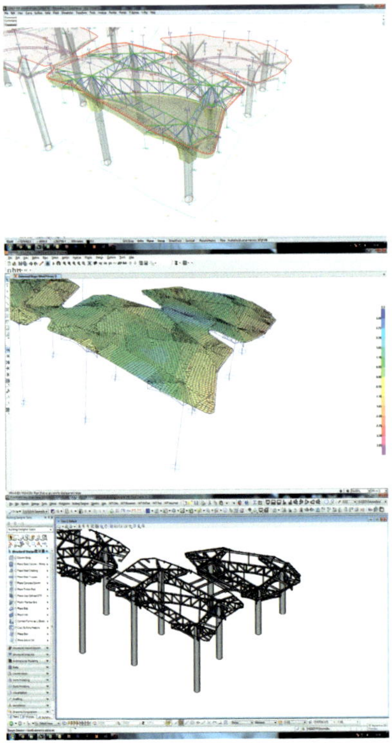

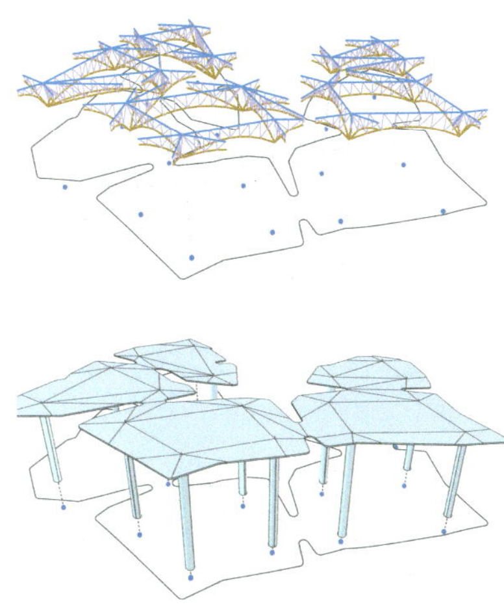

(spherical, patch surface, Catmull–Clark). Optimisation of the crack pattern was explored with the help of the environmental comfort tool, but also by iteratively assessing the seismic effect on the connectivity of the structure under different crack profiles. This constant evolution of the setting-out surface was followed by an updated structural model, and simultaneously by a complete CAD 3D model containing the live structural properties.

MERCHANT SQUARE

The Merchant Square bridge is a perfect illustration of the process flow of designing with Re.AKT. The different parameters influencing the structural behaviour included counterweight size and shape, lever arm, length of hydraulic arms, pivot position, plate thickness, vertical pitch and the tapering depth of deck. Use of the classic linear workflow (sketch to CAD to finite element) would have been highly inefficient in this case, as a modification to any one of the parameters would have had an impact on the others. But by using Re.AKT it was possible to interlink algorithms capable of optimising plate thicknesses, testing the impact that the variation of pivot positions would have on the behaviour of the structure, and influencing the sizes, weights and shapes of counterweights within the extremely tight geometric constraints. So Re.AKT made it possible to define the bridge's geometry, and then vary a set of parameters to achieve the pursued balance between visual quality and structural efficiency. As an example, the non-linear relationship between the counterweight radius and its tonnage had a direct influence on the location of the centre of gravity, which always had to be on the side of the cantilever, keeping the hydraulic rams in compression. This dependency could be varied by either changing the counterweight radius, affecting its visible height above ground, or by changing the location of the pivot, which would have had an impact on the bridge appearance in an open position.

CENTRAL BANK OF IRAQ

This project involved the construction of the new Central Bank of Iraq (CBI) headquarters in Baghdad, adjacent to the River Tigris. Utilising the parametric structural model, multiple scenarios were tested in the geometrical rationalisation exercise which affected the positions of structural fin wall bifurcation. Simultaneously we ran high-level analysis to assess the structure's behaviour in terms of frequency and deflection. The collated data from this exercise allowed us to make an informed decision which ensured changes were not detrimental from a structural aspect, while still satisfying the architectural intent.

The podium roof at the base of the tower was formed by long tentacles joining and parting as they spanned the longer direction across the building. The parametric structural link allowed us to model the internal spatial trusses and determine where the geometry could be altered in a small way to have a considerable influence on the structural interconnectivity of the separate trusses. The architectural vision was maintained while allowing the 60 m spans to be broken into smaller sections. Re.AKT

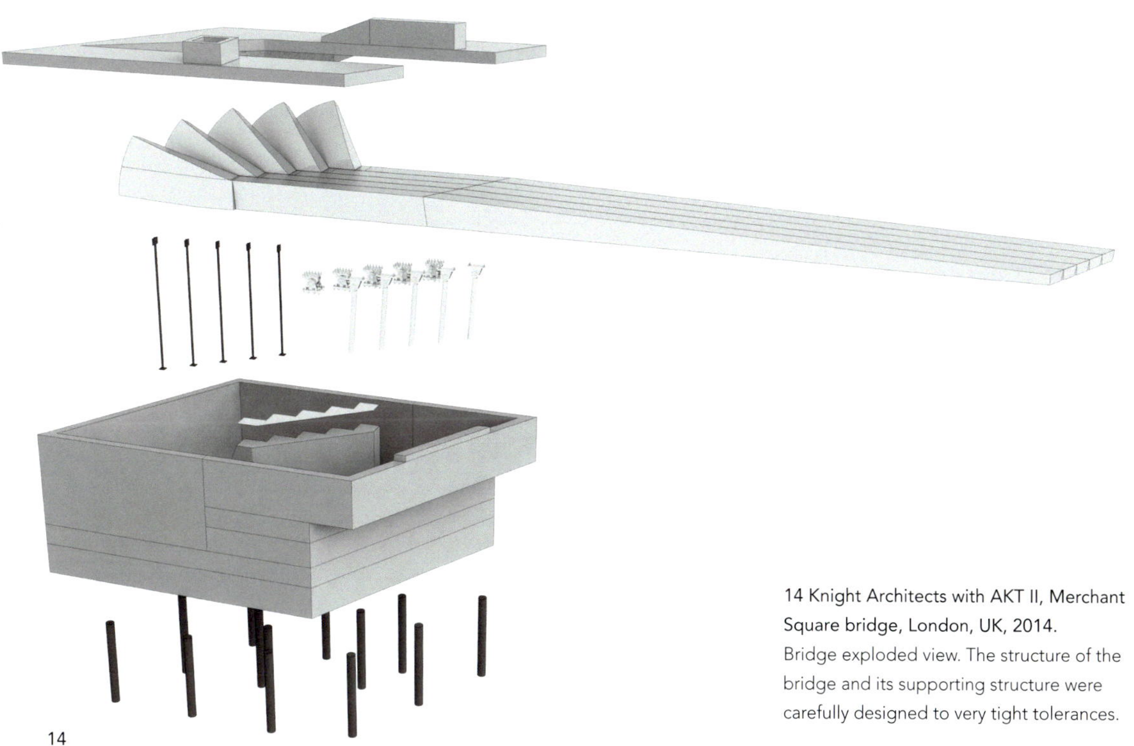

14 Knight Architects with AKT II, Merchant Square bridge, London, UK, 2014.
Bridge exploded view. The structure of the bridge and its supporting structure were carefully designed to very tight tolerances.

15 Knight Architects with AKT II, Merchant Square bridge, London, UK, 2014.
Pivot and piston dependency. The non-linear relationship between the counterweight radius and its tonnage had a direct influence on the location of the centre of gravity.

13 Knight Architects with AKT II, Merchant Square bridge, London, UK, 2014.
Aerial view of bridge in its open configuration.

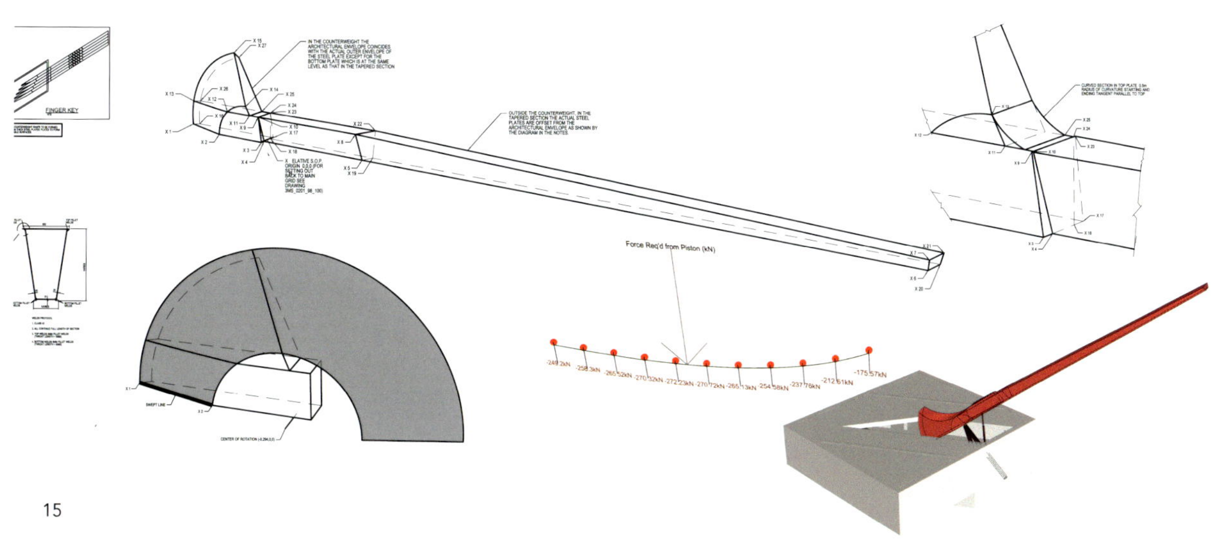

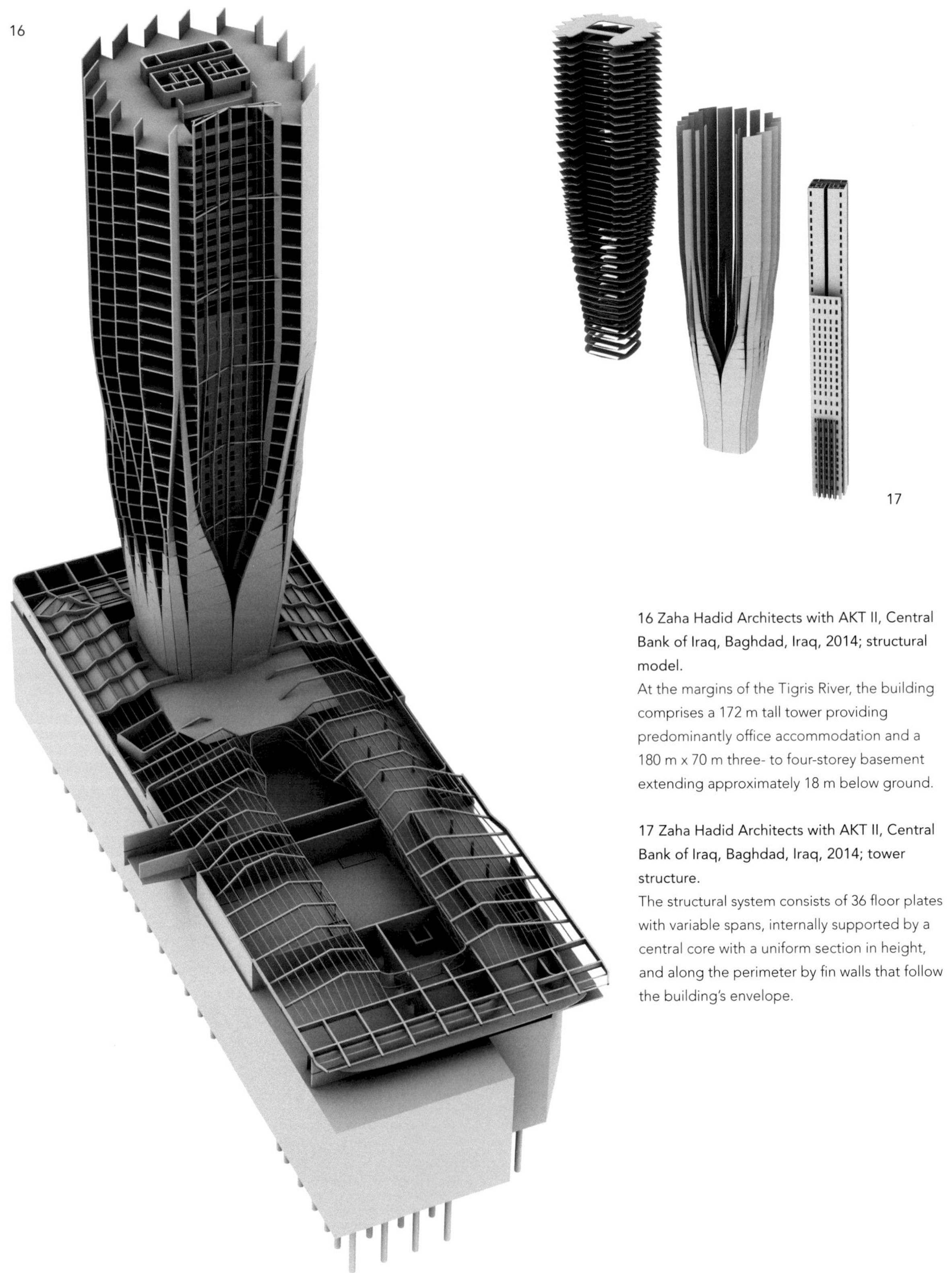

16 Zaha Hadid Architects with AKT II, Central Bank of Iraq, Baghdad, Iraq, 2014; structural model.
At the margins of the Tigris River, the building comprises a 172 m tall tower providing predominantly office accommodation and a 180 m x 70 m three- to four-storey basement extending approximately 18 m below ground.

17 Zaha Hadid Architects with AKT II, Central Bank of Iraq, Baghdad, Iraq, 2014; tower structure.
The structural system consists of 36 floor plates with variable spans, internally supported by a central core with a uniform section in height, and along the perimeter by fin walls that follow the building's envelope.

1 Edge curves

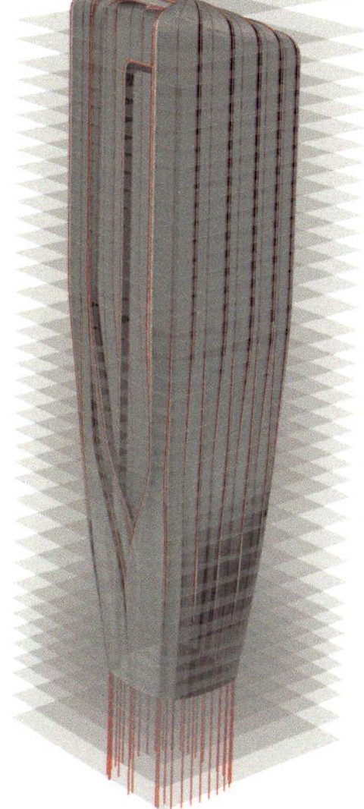

2 TOC Cutting Planes

3 Concrete Wall and
Slab Generation

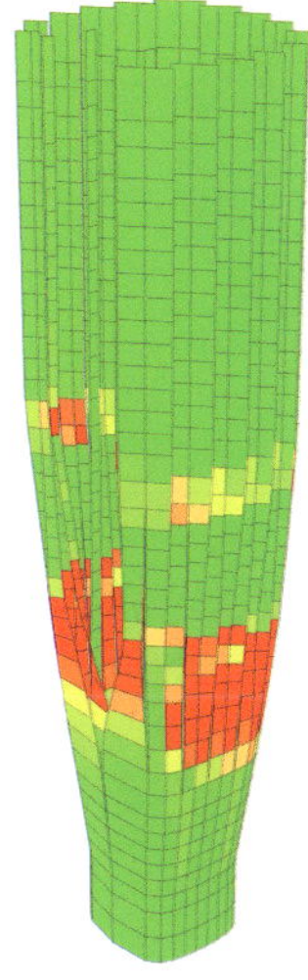

4 Planarity Check

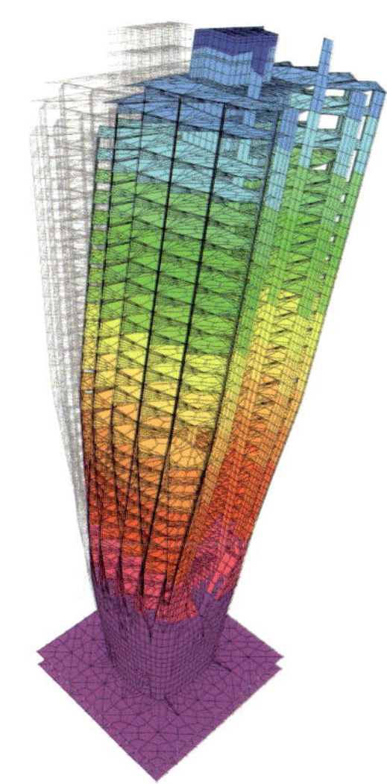

5 Structural Analysis

18

18 Zaha Hadid Architects with AKT II, Central
Bank of Iraq, Baghdad, Iraq, 2014; Re.AKT
design stages.
Boundary surfaces models are turned into
a fully parametric BIM model using Re.AKT,
where finite element method and optimisation
have been run every time the initial geometry
updated.

greatly improved our ability to implement the reinforced-concrete
detailing for construction information; while the visualisation of
the large amounts of data crashed most commercial software, the
reinforcement designer we developed allowed large amounts of
data to be processed to display reinforcement requirements in a
design-friendly and visual way.

The use of in-house custom tools allowed us to visualise and
manipulate colossal amounts of data generated from seismic
and blast analysis, which would have otherwise been a slow and
time-consuming exercise. The custom-visualisation tools not only
created a user-friendly environment to digest the information at
hand, but also helped in making repetitive calculations, as was the
case for the 'staged construction results viewer', where maximum
and minimum differential settlements between the core and fin
walls were automatically calculated and output to a PDF document
for review at every level.

SILO VS HYBRID

Reflecting on the benefits that use of this interface brought to
the AKT II workflow, a clear dichotomy appears: conventional
processes as used in structural engineering so far, with
geometricians, computational teams, engineers and CAD
technicians each working in separate environments, contrasted
with the new process that promoted an active collaboration in a
single environment.

In the first case, the different operators act like silos storing
and manipulating important information regarding the project
in a closed environment, and the task of identifying the crucial
information to share with the rest of design team becomes hard.

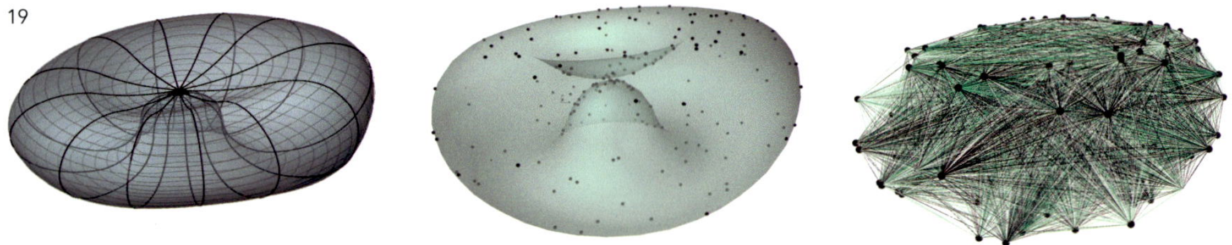

19

19 DeWitt Godfrey with AKT II, Odin, Colgate
University, Hamilton, NY, USA, 2014, Odin
hybrid process.
In Re.AKT the dataset is structured so that the
different parties working on the design are not
progressing their part in isolation, but all the
data collected during the design informs the 3D
model, its shape, its structural behaviour and its
setting out.

20

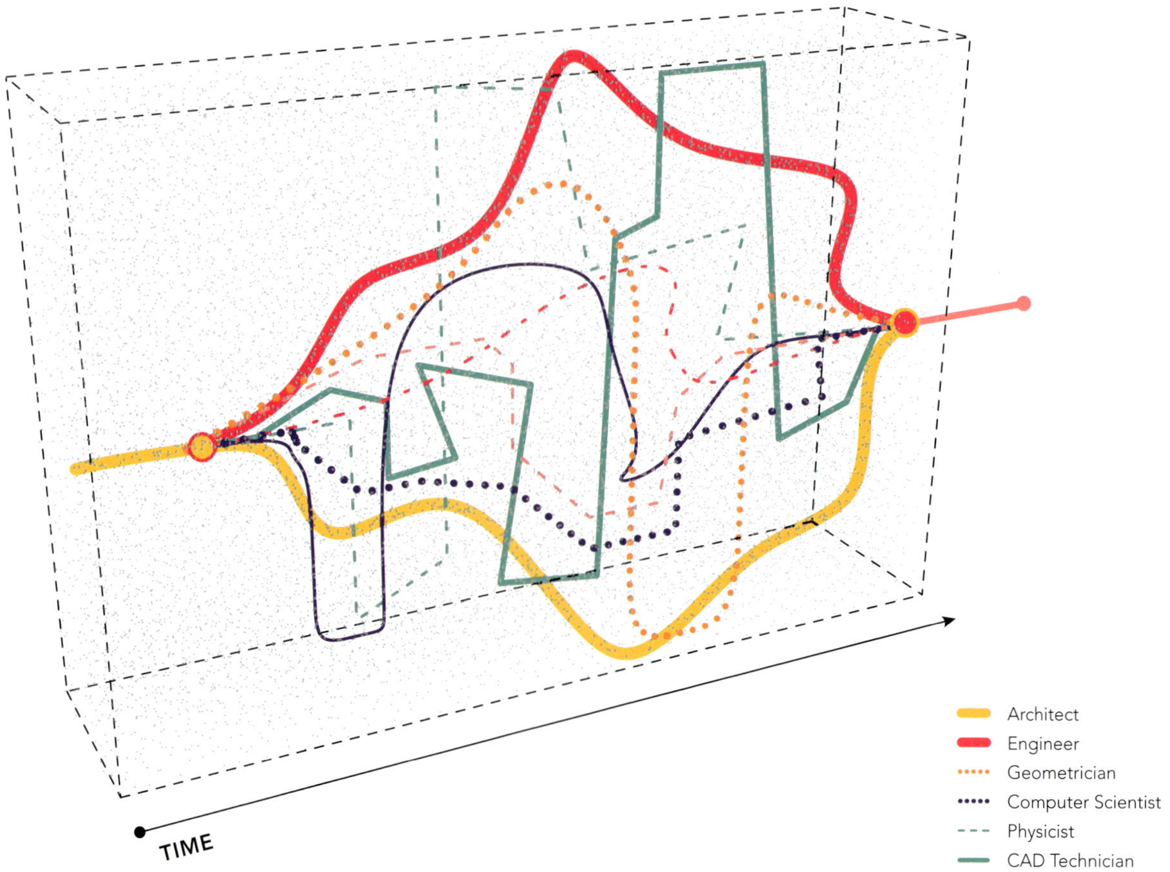

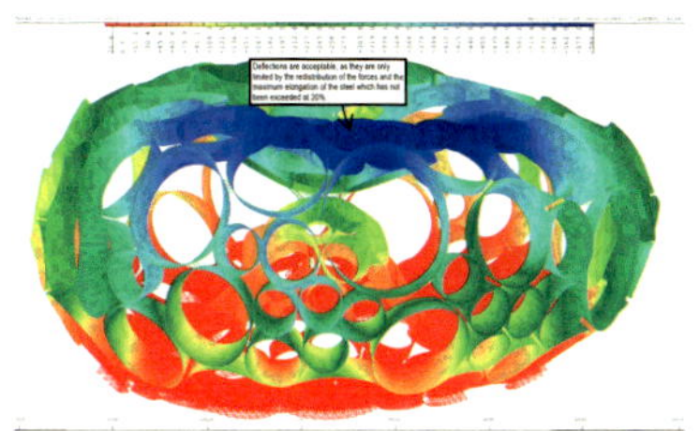

The design suffers from this discrete way of making decisions and, more importantly, an error that is made in one silo will be transferred to the other with all the consequences. Likewise, in this first case, geometricians and CAD technicians, when engaged in a conventional project, usually start producing information in a form of 2D linear drawings, only to move to the 3D environment in the later stages of the project or when required by a specific contractual clause.

In contrast, we have tried with Re.AKT to enhance the connectivity between the different parties involved in the design, thus creating a complete overlap between engineers, geometricians, CAD technicians, computer scientists and structural analysts in a new 'hybrid' ecosystem.

We wanted to create an environment where at any point the dataset could be improved and augmented, not only internally, but also by decisions taken by the wider design team. This allowed the team to react to any evolution of the project. No longer acting in isolation, spending energy on converting data from one package to another, they could now concentrate as a unit on the challenges of the structural limits rather than on the software limitations.

Re.AKT is an experiment in creating a collaborative process, both internally and externally. The operator that works in this environment must have not only an engineering background and CAD technician training with in-depth knowledge of geometry, but also an architectural background which enables an understanding of the requirements of the project. The evolution of the designer is resulting in an individual who has a competence in all those fields. A new character will emerge, an über-designer who is experienced enough to anticipate the architect/client needs and engage with a fast responsive tool that can sift many iterations, and produce solutions that are efficient, cost-effective and visually pleasing, as well as code compliant and comfort compatible.

THE BEYOND

With a broad, collaborative and open perspective, design engineers have engaged in a deeper dialogue with architects, other consultants and fabricators, allowing creativeness and rigour to merge in a complex and iterative process of design. Their ability to develop tools that allow for the interaction of human intuition

20 AKT II, 2015; enhancing professional figure. As the progress contributed to delaminating the practice of both Architecture and Engineering with the rise of specialist disciplines, the use of an inclusive workflow that needs data from all those disciplines to improve the design could result in a rapprochement of the practices.

with analytical computer data processes, has led to a preference for complex hybrid solutions, rather than more reductive and self-referential purist extremes.

As part of a fast ongoing transformation of the disciplines of design, we can only speculate on its future outcomes, but we can be optimistic, considering the much deeper reciprocal interest of the two fields in each other's values and struggles. We share the same economic, political and environmental challenges and a collaborative approach seems the best way forward.

By embracing technological advancement, with a forensic approach to design, we advocate creativity. Starting from the fundamental understanding of geometry, materiality, the detailed behaviour of structures and their environment, we create a springboard for innovation. We may be at the emergence of a new era of humanism, where technical excellence and intellectual enlightenment will pave the way for a better world.

REFERENCES

1 Ivan Sutherland. *Sketchpad: A Man-Machine Graphical Communication System*, MIT doctoral thesis, 1963.

2 3D ACIS Modeler is a geometric modelling kernel developed by Spatial Corporation (formerly Spatial Technology), part of Dassault Systèmes. ACIS is used by many software developers in industries such as computer-aided design (CAD), computer-aided manufacturing (CAM), computer-aided engineering (CAE), architecture, engineering and construction (AEC), coordinate-measuring machine (CMM), 3D animation, and shipbuilding. ACIS provides software developers and manufacturers with the underlying 3D modelling functionality. (Wikipedia, https://en.wikipedia.org/wiki/ACIS)

3 AutoCAD is a commercial software application for 2D and 3D computer-aided design and drafting, available since 1982 as a desktop application and since 2010 as a mobile web- and cloud-based app marketed as AutoCAD 360. (Wikipedia, https://en.wikipedia.org/wiki/AutoCAD)

4 Building information modelling (BIM) is a process involving the generation and management of digital representations of physical and functional characteristics of places. Building information models (BIMs) are files (often but not always in proprietary formats and containing proprietary data) which can be exchanged or networked to support decision-making about a place. (Wikipedia, https://en.wikipedia.org/wiki/Building_information_modeling)

5 Autodesk Revit is building information modelling software for architects, structural engineers, MEP engineers, designers and contractors. It allows users to design a building and structure and its components in 3D, annotate the model with 2D drafting elements, and access building information from the building model's database. Revit is 4D BIM capable with tools to plan and track

various stages in the building's life cycle, from concept to construction and later demolition. (Wikipedia, https://en.wikipedia.org/wiki/Autodesk_Revit)

6 Finite element analysis (FEA) and computational fluid dynamics (CFD).

7 Application programming interface (API) is a set of routines, protocols and tools for building software and applications. An API expresses a software component in terms of its operations, inputs, outputs and underlying types, defining functionalities that are independent of their respective implementations, which allows definitions and implementations to vary without compromising the interface. A good API makes it easier to develop a program by providing all the building blocks, which are then put together by the programmer. (Wikipedia, https://en.wikipedia.org/wiki/Application_programming_interface)

8 i-model is an interoperable model interface which allows the tagging of properties to a parametric/geometric model for the real-time computation of structural and environmental results. The interface is designed to allow the feedback of analysis results into the parametric design model, through the use of automated multi-parametric routines or interactive graphic tools.

9 CATIA (an acronym of computer aided three-dimensional interactive application) is a multi-platform CAD/CAM/CAE software suite developed by the French company Dassault Systèmes. (Wikipedia https://en.wikipedia.org/wiki/CATIA)

10 Digital Project is a computer-aided design (CAD) software application based on CATIA V5 and developed by Gehry Technologies, a technology company owned by the architect Frank Gehry. Among the changes made by Gehry Technologies to CATIA is a new visual interface suitable for architecture work

TEXT

IMAGES

EDITOR BIOGRAPHIES

PROFESSOR HANIF KARA

Hanif Kara is a London-based structural engineer and educator. Since co-founding Adams Kara Taylor his particular interests in innovative form, material uses and complex analysis methods have allowed him to work on award-winning and unique projects with leading designers. These include Peckham Library in the UK, Phæno Science Centre in Germany and the Masdar Institute in Abu Dhabi.

In 2008 he accepted a position as visiting Professor of Architectural Technology at KTH in Stockholm, Sweden and is currently Professor in Practice of Architectural Technology at the Graduate School of Design, Harvard.

His approach extends beyond the technical aspects of structural engineering and led to his appointment as a commissioner for the Commission for Architecture and the Built Environment (CABE) in 2008. In 2004 he was appointed to the Master Jury of the Aga Khan Awards for Architecture where he is currently serving as member of the Steering Committee for the AKAA. He is also a fellow of the RAE, RIBA, ICE and IStructE, on the board of trustees of the Architecture Foundation and a member of the BCO Technical Affairs Committee.

He has published and widely lectured on the subject of design. Notable publications include *Design Engineering* (Actar, 2008) and *Interdisciplinary Design: New Lessons from Architecture and Engineering* (Kara/Georgoulias, Actar/Harvard, 2012).

DANIEL BOSIA

Director at AKT II and head of the specialist team p.art, Daniel is a qualified structural engineer with an MSc in Structural and Bridge Engineering and a Masters degree in Architecture. He is an expert in computational and structural design.

Formerly head of the AGU at Arup, Daniel has over 18 years of experience working on complex, high-profile multidisciplinary projects including buildings, footbridges and large-scale art installations. Daniel worked on the Serpentine Pavilion in 2002 and the Taichung Opera House with Toyo Ito & Associates, the New Scottish Parliament Building with Enric Miralles and Benedetta Tagliabue, the Frankfurt Messehalle with Grimshaw and the Arnhem Centraal station with UN Studio. More recently he has been involved in the design of the Bloomberg Headquarters with Foster + Partners and the Google Headquarters in Mountain View with BIG and Heatherwick Studio.

Daniel holds a teaching position at Columbia University after running a Diploma Unit at the Architectural Association in London, and is Honorary Professor at the Department of Architecture, Design and Media Technology in Aalborg. He has lectured at architectural and engineering schools in the US and Europe including UPenn, IIT, Princeton, Yale, Harvard, ETH and UCL Bartlett.

CONTRIBUTOR BIOGRAPHIES

JORDAN BRANDT is a tech entrepreneur, investor and consulting Associate Professor of Engineering at Stanford University. He was formerly a member of p.art at AKT II.

MARCO CERINI studied Structural Engineering at Politecnico di Milano then went on to complete a doctorate at Imperial College London. He is now a senior engineer at AKT II.

DIEGO CERVERA DE LA ROSA studied Architecture and a Masters in Structural Engineering at the Technical University of Madrid. He is a senior engineer at AKT II.

DR PHILIP ISAAC studied Engineering at the University of Bath, where he also holds a visiting research fellowship. He is a design engineer at AKT II.

JEROEN JANSSEN studied Architecture at Eindhoven University of Technology. He is a core member of p.art at AKT II.

SAWAKO KAIJIMA is an Assistant Professor of Architecture and Sustainable Design at the Singapore University of Technology and Design. She previously led the computational design team at AKT II.

JAMES KINGMAN studied Civil and Structural Engineering at the University of Leeds. He is a design engineer at AKT II.

ALESSANDRO MARGNELLI studied Civil Engineering at the University of Rome Tor Vergata. He is an associate director at AKT II.

PANAGIOTIS MICHALATOS is Associate Professor of Architecture at Harvard GSD, and previously led the computational design team at AKT II.

ED MOSELEY studied Civil Engineering at the University of Warwick. He is a design director within AKT II and takes an active role in academia and research.

RICHARD PARKER studied Civil and Structural Engineering at the University of Sheffield. He is a senior structural engineer at AKT II.

ANDREW RUCK studied Civil Engineering with European Studies at the University of Bristol. He is a design director at AKT II.

ADIAM SERTZU studied Architecture at the AA School of Architecture. She is an associate at AKT II, in p.art.

DJORDJE STOJANOVIC is the founder of 4of7 Architecture and an Assistant Professor at the University of Belgrade. He was formerly head of p.art at AKT II.

EDOARDO TIBUZZI studied Civil Engineering at the University of Rome Tor Vergata. He is a designer and researcher, and currently team leader of AKT II's computational unit.

DR MARTIJN VELTKAMP is trained as a structural engineer and is currently head of the engineering department at the 'design & build' firm, FiberCore in the Netherlands. He was previously a senior engineer at AKT II.

MARC ZANCHETTA is an aeronautical engineer with a passion for the built environment, and is founder of GasDynamics Ltd, which specialises in aerodynamics and thermodynamics. He has collaborated with AKT II on several projects.

SELECT BIBLIOGRAPHY

Journal of Architectural Education, Vol 62/3, Feb 2009, Avigail Sachs, p 53.

Aldo Capasso and Frei Otto, *Le Tensostrutture a membrana per l'architettura*, Maggioli Editore (Rimini), 1993.

Carlos Felippa, *Introduction to Finite Element Methods* (ASEN 5007), Department of Aerospace Engineering Sciences (University of Colorado, Boulder), 2006.

Lisa Iwamoto, *Digital Fabrications: Architectural and Material Techniques*, Princeton Architectural Press (New York), 2009.

Jeroen Janssen and Adiam Sertzu, *Addis Ababa: Experiment on Building Envelopes: Digital Vernacular*, AKT II (London), 2014.

Branko Kolarevic, 'Digital Fabrication: Manufacturing Architecture in the Information Age', University of Pennsylvania.

Rafael Sacks and Rebecca Partouche, 'Production Flow in the Construction of Tall Buildings', ASCE Construction Research Congress, 2009.

INDEX